Silverflexagons

and the Flexatube

Fascinating flexible novelties and puzzles made from folded strips of paper

David Mitchell

A Water Trade publication
www.watertradebooks.co.uk
Copyright David Mitchell 2015
All rights reserved

About Water Trade

Water Trade is a micro-publisher of books about aspects of origami and paperfolding. We aim to publish books containing original material of the highest quality organised around technically interesting or unusual themes, and which are consequently far more than simply collections of instructions for designs. We welcome submissions from new or established authors on potential themes for future books.

About David Mitchell

David Mitchell has been a professional author, illustrator and designer specialising in origami for many years. He lives in Cumbria, close to the mountains of the English Lake District, where he loves to walk. He is also a passionate fan of latin dance and music, particularly salsa and bachata.

As a designer he is particularly well known for his innovative modular and macromodular sculptures but is also a prolific inventor of single and multiple piece paperfolds, action novelties and puzzles.

His other books include Complete Origami, Mathematical Origami, The Magic of Flexagons, Paper Crystals, Origami Animals, Building with Butterflies, Paperfolding Puzzles, Origami Alfresco, Paper Planes, and Sticky Note Origami.

David Mitchell can be contacted through his website www.origamiheaven.com

Contents

Introduction	7
Glossary of technical words	9
Key to the flexigation symbols	10
The Zigzag Silverflexagon	13
The Woven Flexatube	27
Slit-Square Silverflexagons	37
The First Slit-Square Silverflexagon	38
Non-Twisted First Slit-Square Silverflexagons	47
Making the First Slit-Square Silverflexagon from a single square	54
Other Slit-Square Silverflexagons	57
The Labyrinth Silverflexagon	59
The Flexatube	85
Compound Flexatubes	89
Flexatube Stack Puzzles	90
Flexatube Chain Puzzles	92
The Flexamat Puzzle	93
Solutions to the Flexatube	95
Solutions to the Compound Flexatubes	115
Templates	127

Introduction

This book combines material about silverflexagons previously published in the Water Trade leaflets Silverflexagons 1, 2 and 3 in 2002 and about the Flexatube previously published in the first edition of Paperfolding Puzzles in 1998 (but which was omitted from the second edition so that it could be republished here). The material on Compound Flexatubes is new to this volume.

About flexagons

It is not the purpose of this book to provide a general introduction to the subject of flexagons but some brief notes and definitions may be helpful. A pdf entitled Flexagons and Flexigation, a more detailed introduction to flexagons, which explains how some simple tetraflexagons can be flexed and the principles by which they can be mapped, can be downloaded without charge from the author's website www.origamiheaven.com.

Since there is no widely agreed definition of what a flexagon is perhaps the best way to explain them is to say how they are made and how they behave. To make a flexagon you take a strip of paper, either straight or of some convoluted shape, divide it into a number of segments, usually of the same shape, by creases, fold it up / weave it into a polygonal shape and join the ends of the strip together. In the process some of the surfaces of some of the segments will have been hidden inside the weave. If the result is a flexagon it will now be possible to flex it along the lines of the creases between the segments so that the faces of the segments hidden inside the weave can be brought into view.

The behaviour of a flexagon largely depends on the shape of the segments the strip is divided into because this determines the angles at which the creases between them are set. Or vice versa, of course, depending on your point of view. Flexagons whose segments are rectangular are called tetraflexagons. Flexagons whose segments are equilateral triangles are known as hexaflexagons. You will be able to find lots of material on the net and in other books about these two types of flexagon.

About silverflexagons

Silverflexagons are less well known. They are made from strips divided into segments in the shape of right angle isosceles triangles. I call such triangles 'silver triangles' because they share with the silver rectangle (1:sqrt2) the property that they can be bisected into similar shapes. The word silverflexagon is my own invention derived from this usage. Les Pook, the author of Serious Fun with Flexagons, a fascinating and encyclopaedic book that unfortunately only university libraries and true enthusiasts will probably afford to buy, has adopted my usage but separates the words, thus, silver flexagon. In this book I have still, however, preferred to combine the two words into one in keeping with the established usages of hexaflexagon and tetraflexagon.

Silverflexagons are more complex than their better known cousins and as a result they are more interesting to explore and analyse, one of their most interesting properties being that they change shape from square to rectangular to hexagonal to pentagonal as they are flexed. In mapping the silverflexagons explained in this book it has been necessary to decide which of these shapes constitute states of the flexagon and which are just intermediate forms that occur during the flexing process. A mathematical analysis of silverflexagons might provide a definitive answer, but unfortunately one does not yet exist. In the meantime I have done the best I can on a common sense basis.

A word about language

In writing about flexagons I have often found myself at a loss for an established word to describe the flexagons themselves, their constituent parts and the way they can be flexed. In addition I have found that some of the existing language is unhelpful. So, where I needed new, or better, words, I have quite unashamedly invented them. There is a short glossary on the next page which will, I hope, explain my usages clearly.

Why include the Flexatube?

The Flexatube is not a silverflexagon, nor indeed, any kind of flexagon at all, but it is very like one. It is made from the same strip of silver triangles that will make the Woven Flexatube and although it does not have hidden faces, and can only be folded not flexed, it behaves, at times, in a very similar way. At other times it behaves very differently, of course. Above all the Flexatube is, like the silverflexagons themselves, immense fun to play with and explore. And for that alone it deserves its place.

About the Templates

Templates for each of the flexagon and Flexatube designs are included in this book. These are intended to be scanned and printed out, or photocopied, so that you can easily make the designs to play with and explore for yourself. Because of the format of this book these templates are, however, smaller than they would ideally have been, so you will also find larger, more colourful, downloadable templates in pdf format on the author's website www.origamiheaven.com.

About complexity and chaos

As flexagons get more complex they also tend to become more chaotic, which is to say that there are some points during some flexes at which the stacks may accidentally rearrange themselves in unexpected ways, or that you might make a flex you were not quite expecting to make, with the result that you become lost in a kind of flexagon fog. This is one reason why the maps are so detailed. By comparing the flexagon in your hand to the maps you should always be able to find your way back to the home position.

And finally

At the end of the day flexagons are not really about language, flexing theory or maps, however interesting all these may be. They are about the sheer joy of holding one in your hand and flexing it. Enjoy!

Glossary of technical words

Bronzeflexagon: A flexagon whose segments are 1:2:sqrt3 triangles.

Chaotic: Descriptive of an unstable configuration of a flexagon in which the stacks may become rearranged in an unintended fashion.

Extended flexagon: A flexagon in which all the segments do not lie completely across the strip from edge to edge, for instance, the Woven Flexatube.

Faces: The visible surfaces of a state, composed of the visible surfaces of a number of segments. Each state will have two faces.

Flex: A manoeuvre in which the flexagon is first folded along one of more complete or partial lines of symmetry through the focus and then opened out with the effect of moving segments between stacks and thus transforming the flexagon into a new state.

False flex: A manoeuvre in which segments are moved between stacks without the flexagon being folded in half and opened out.

Focus: The centre, or centres, of symmetry of a state. A focus may be said to be closed when the segments meet in the centre and open when they do not.

Home state: The state or states of the flexagon where the segments are arranged most symmetrically around the focus.

Non-twisted flexagon: A flexagon made from a strip which can be cut complete from a flat sheet of paper (and which does not, therefore, need the two ends of the strip glueing together as the flexagon is constructed).

Pure flexagon: A flexagon in which all the segments lie completely across the strip from edge to edge, for instance the Zigzag Flexagon.

Segment: A division of the paper strip from which the flexagon is made bounded by creases along which the strip will fold. Segments are sometimes also known as leaves.

Stack: A pile of segments which lie on top of each other in a state.

State: A stable arrangement of the flexagon which may be flat or pyramidal depending on the number of stacks in the flexagon.

Key to the flexigation symbols

Each of the flexes explained in this book is shown on the maps using a specific symbol.

Rolling flex: Fold the flexagon in half edge to edge then open out in the same direction. For visual explanation see page 18.

Tube flex: A rolling flex in which the stacks are rearranged by first forming, then flattening in the alternate direction, a square section tube in between the two halves of the rolling flex. For visual explanation see page 19.

Diagonal tube flex: A tube flex made in a diagonal direction. For visual explanation see pages 30 and 31.

Single tuck flex: A rolling flex made with one corner tucked in. For visual explanation see page 23.

Double tuck flex: A rolling flex made with two corners tucked in. For visual explanation see page 24.

Open single tuck flex: A single tuck flex made without folding the flexagon in half. Open tuck flexes change one hexagonal or pentagonal state into another of the same shape. For visual explanation see page 25.

Double flip flex: A diagonal flex in which the stacks are rearranged by flipping the corners inside out. For visual explanation see pages 15, 16 and 17.

Single swivel flex: A diagonal flex in which the stacks are rearranged by swivelling a single corner.

Double swivel flex: A diagonal flex in which the stacks are rearranged by swivelling two corners. For visual explanation see pages 21 and 22.

Scenic route: A complex diagonal flex via a two-pocket parallelogram. For visual explanation see page 32, 33 and 34.

 Swivel/flip flex: A diagonal flex that is half swivel flex and half flip flex. The flat end of the central symbol represents the swivel flex and the indented end the flip flex. For visual explanation see page 46.

 Pop-up swivel flex: A false flex that begins by pulling the focus apart. For visual explanation see pages 50 and 51.

 Inside-out flex: A complex double tuck flex which turns the flexagon completely inside out. For visual explanation see pages 63 and 64.

 Open squash flex: An open flex which changes a pentagonal state into a hexagonal one and vice versa. For visual explanation see page 70.

 Open squash flex: An open flex which changes a pentagonal state into a rectangular one and vice versa. For visual explanation see pages 71 and 72.

 Primary sub-level access flex: A complex double tuck and inside swivel flex which gives access to the primary sub-levels of the Labyrinth Silverflexagon. For visual explanation see pages 72 and 73.

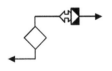 Secondary sub-level access flex: A complex double tuck, inside swivel and tube flex which gives access to the secondary sub-levels of the Labyrinth Silverflexagon. For visual explanation see pages 76 and 77.

 Tertiary sub-level access flex: A complex single tuck, inside swivel and diagonal flex which gives access to the tertiary sub-levels of the Labyrinth Silverflexagon. For visual explanation see pages 80 and 81.

The Zigzag Silverflexagon

The Zigzag Silverflexagon

The Zigzag Silverflexagon is made from a strip of twelve silver triangles arranged in the form of a zigzag. It can be flexed into twelve flat states, of which four are square, two oblong (2x1) and eight hexagonal. The Zigzag Silverflexagon can be flexed in at least eight different ways.

Making the Zigzag Silverflexagon

A template for this flexagon can be found on page 128.

1

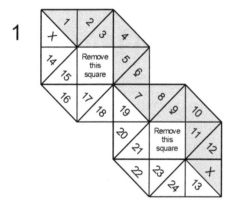

2

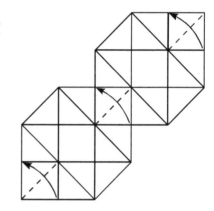

1. Cut the template out and crease carefully along all the boundaries between the segments. Fold all the creases backwards as well as forwards so that the segments move freely in both directions.

2. Apply glue to one half of the plain side of the template then fold one half onto the other so that all the edges line up. If necessary, trim slightly to neaten the edges.

3

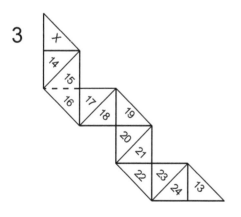

4

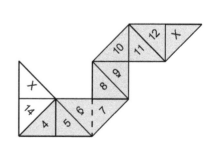

3. Arrange the strip like this and fold segment 16 onto segment 15.

4. Now fold 7 onto 6.

David Mitchell / Silverflexagons and the Flexatube

5 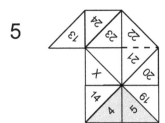 6 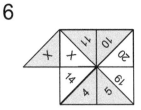 7

5. Fold 22 onto 21 so that 24 goes beneath segment x.

6. Fold and glue X onto X.

7. This is state A of the Zigzag Silverflexagon.

A mirror-image version of the Zigzag Silverflexagon can be made by reversing the direction of all the folds.

Flip flexes

Flip flexes are diagonal flexes in which the stacks are rearranged by flipping the corners inside out. The double flip flex explained here is the quickest route between the two square states of the Zigzag Silverflexagon.

8 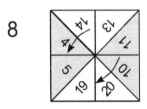 9 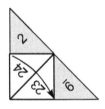 10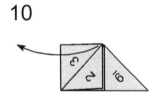

8. Fold 14 onto 4 and 10 onto 20.

9. Fold 24 onto 23 so that 2 and 3 become visible.

10. Fold 3 across to the left.

11 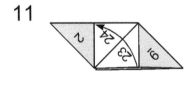 12 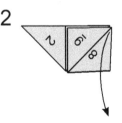 13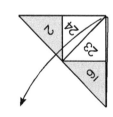

11. Fold 23 onto 24 so that 8 and 9 become visible.

12. Fold 8 downwards.

13. Fold the front layers diagonally downwards and to the left.

David Mitchell / Silverflexagons and the Flexatube

14

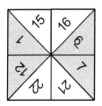

14. This is state B of the Zigzag Silverflexagon.

Diagonal double flip flex map of the Zigzag Silverflexagon

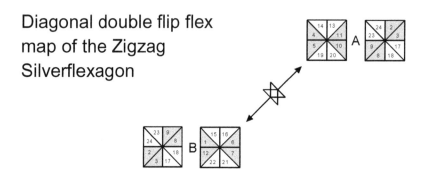

States A and B are the home positions of the Zigzag Silverflexagon. You will notice that each surface of every segment occurs once, and only once, in one of the four faces.

The flexagon maps and flexing instructions in this publication are drawn as if the flexes are being made while the flexagon is lying flat on a tabletop but this is not always the best way to perform the flexes when you are holding the flexagon in your hands.

A double flip flex between states A and B can also be performed, for instance, perhaps more naturally, in the way shown below. Begin by following instruction 8 then continue with 15 through 20 below.

15 **16** **17**

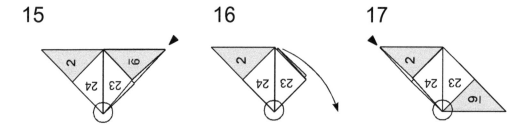

15. Hold the flexagon gently between the thumb and forefinger of your left hand at the point marked with a circle. Push the top right hand corner inside out between the layers.

16. Separate the layers at the top centre and flip the free point downwards to the right.

17. This is the result. Change hands. Repeat step 15 on the left hand side of the flexagon.

18 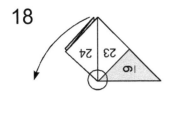 **19** 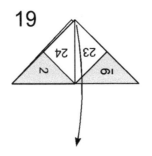 **20**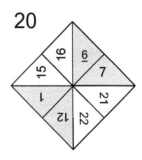

18. Repeat step 16 on the left hand side.

19. Fold the front layers downwards.

20. This is state B.

Pinch flexes

The term 'pinch flex' will already be familiar to anyone who has studied hexaflexagons. When pinch flexing a silverflexagon you need to begin the flex by bringing four corners, rather than three in the case of hexaflexagons, together underneath the focus. Pinch flexes are just double flip flexes in disguise so do not need to be mapped separately.

21 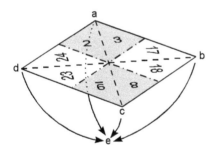 **22**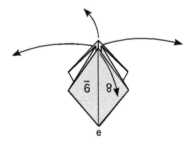

21. Begin with state A. Fold corners a, b, c and d downwards so that they meet at point e.

22. Separate the corners at the top and pull them apart.

23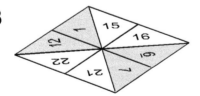

23. This is state B. To return to state A turn the flexagon over and repeat the same set of moves.

David Mitchell / Silverflexagons and the Flexatube

Rolling flexes

Rolling flexes are the most fundamental type of flex. Here's how to make one.

24 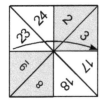 25 26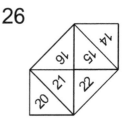

24. Begin with state A. Fold in half sideways to the right.

25. Open out the front layers by folding them across to the right.

26. This is state E.

A rolling flex map of the Zigzag Silverflexagon is shown below. You will notice that square states A and B are not directly linked by rolling flexes.

Rolling flex map of the Zigzag Silverflexagon

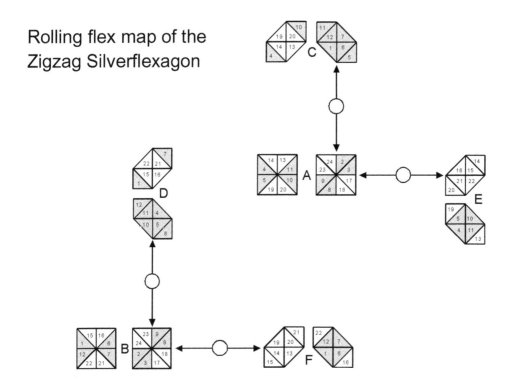

You will also notice that the faces of states C, D, E and F are all hexagonal rather than square. It is a general (though not universal) property of silverflexagons that a rolling flex changes a square face into a hexagonal one.

Tube flexes

If you try to move between states E and F by making a rolling flex in a downwards direction from the lower face of state E (see Rolling flex map) you will find that it is not possible to do so. These states can however be linked by making a tube flex in the way shown below.

 27 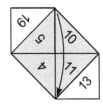 28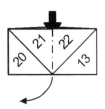

28. Separate the centre to create a square section tube.

27. Begin with state E. Fold 5/10 onto 4/11.

29 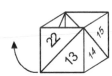 30 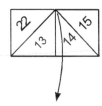 31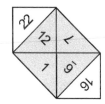

29. Flatten the tube in the alternate position.

30. Fold the front layers downwards.

31. The Zigzag Silverflexagon is now in state F.

Tube flex map of the Zigzag Silverflexagon

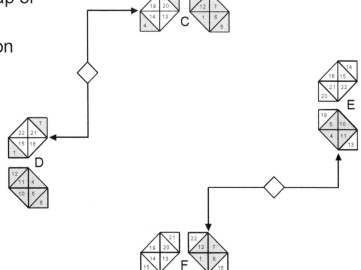

David Mitchell / Silverflexagons and the Flexatube

It is also possible to move between states A and B using tube flexes like this.

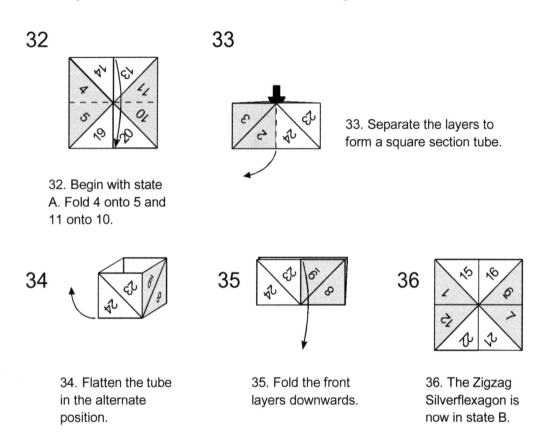

32. Begin with state A. Fold 4 onto 5 and 11 onto 10.

33. Separate the layers to form a square section tube.

34. Flatten the tube in the alternate position.

35. Fold the front layers downwards.

36. The Zigzag Silverflexagon is now in state B.

This is how this flex would appear on a map

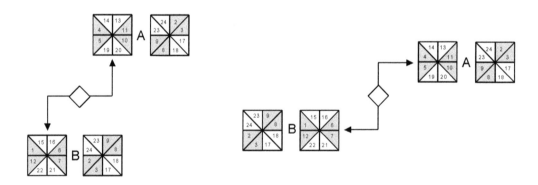

Because of the symmetry of states A and B it is also possible to arrive at the same result in another way.

Rolling flexes and tube flexes are closely related and can be combined on a single map like this.

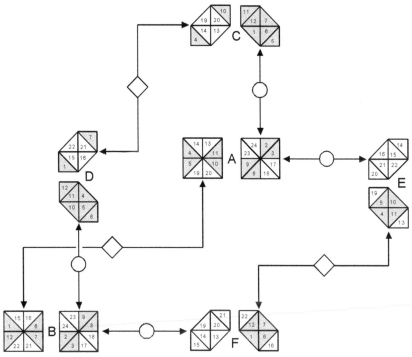

Double swivel flexes

Swivel flexes are a type of diagonal flex in which the stacks are rearranged by swivelling flaps of segments released by the diagonal fold. A flex in which two such flaps are released and swivelled is a double swivel flex.

37

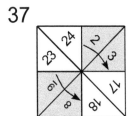

37. Begin with state A. Fold 2 onto 3 and 9 onto 8.

38

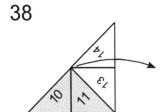

38. You will find that segments 14 and 13 form a loose flap. Swivel this to the right as shown.

39

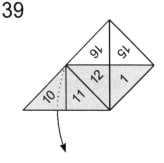

39. There is another loose flap behind segments 10 and 11. Swivel this into view in a similar way.

40

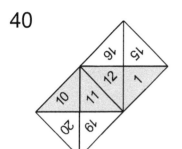

40. This is state M. Turn over sideways.

41

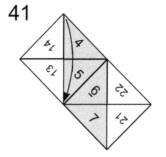

41. Fold 14/4 onto 13/5.

42

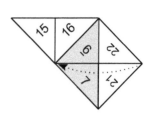

42. Fold 22/21 behind 6/7.

43

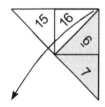

43. Fold the front layers diagonally downwards to the left.

44

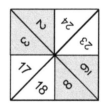

44. This is state B.

The Zigzag Silverflexagon has two oblong states. Not every surface of each segment occurs in an oblong face.

Double swivel flex map of the Zigzag Silverflexagon

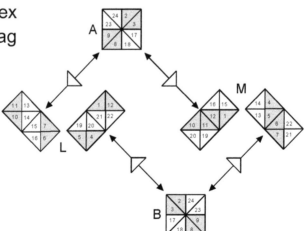

22   David Mitchell / Silverflexagons and the Flexatube

Tuck flexes

Tuck flexes are a type of rolling flex. They also turn square faces into hexagonal ones.

45

46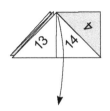

47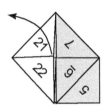

45. Begin with state A. Fold 9 onto 8 then 2 onto 18.

46. Fold the front two layers downwards.

47. Open out flap 21/22 as shown.

48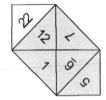

49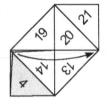

50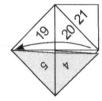

48. This is state K. Turn the flexagon over sideways.

49. Fold 14 onto 13.

50. Fold 20 onto 19.

51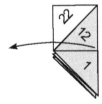

52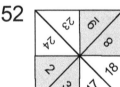

51. Fold the front layers across to the left.

52. This is state B.

The map overleaf shows the four hexagonal faces that can be discovered by performing single tuck flexes in this way. The black triangles on the symbols used to represent the flexes tell you which corner should be tucked in as the flex is performed.

You will note that state B appears twice on this map. A similar map could be drawn with state B at the centre and state A shown twice.

Single tuck flex map of the Zigzag Silverflexagon

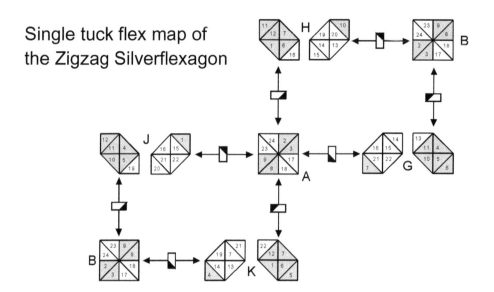

It is also possible to perform double tuck flexes like this.

53 **54** **55** **56**

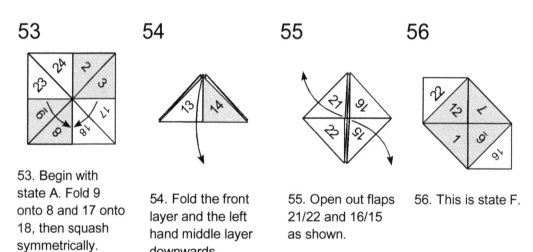

53. Begin with state A. Fold 9 onto 8 and 17 onto 18, then squash symmetrically.

54. Fold the front layer and the left hand middle layer downwards.

55. Open out flaps 21/22 and 16/15 as shown.

56. This is state F.

The maps show that it is possible to flex directly from state A to states D and F and from state B directly to states C and E.

Double tuck flex maps of the Zigzag Silverflexagon

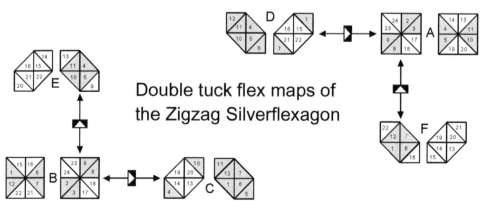

Single Open Tuck flexes

An open flex is a type of false flex that can be performed while the flexagon is open (as opposed to a rolling flex or a diagonal flex in which the first part of the flexing procedure is to close the flexagon by folding it in half).

Using single open tuck flexes hexagonal states C and F can be transformed directly into hexagonal states H and K (and consequently also into each other) without going through square states A or B. Similarly, states D and E can be transformed directly into states G and J (and consequently into each other).

A single open tuck flex only re-orders the stacks at one end of the flexagon. Its effect is to transfer the loose flap (look at segments 10 and 20 in picture 57 below) from one face to the other. Because of this, turning the flexagon over and performing the identical moves on the same end of the flexagon restores the flexagon to its previous state. The way these flexes have been mapped, using one way arrows, reflects this property (although the flex can, of course, be reversed without turning the flexagon over).

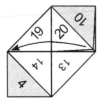

57. This is state C. Fold 20 onto 19.

58. Fold 10 onto 11.

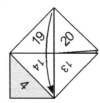

59. Fold 19/20 onto 14/13

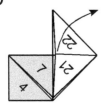

60. Open out flap 22/21 as shown.

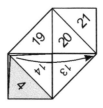

61. This is state K. To flex to state F make another single open tuck flex on the opposite corner.

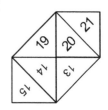

62. This is state F.

David Mitchell / Silverflexagons and the Flexatube

These maps show that there are two independent cycles of single open tuck flexes. The black triangles on the symbol used to represent the flexes tell you which corner should be tucked as the flex is made.

Single open tuck flex maps of the Zigzag Silverflexagon

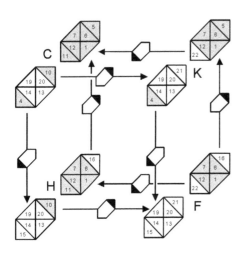

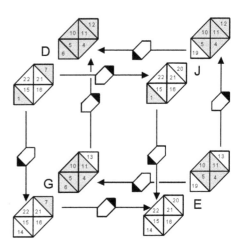

This concludes our exploration of the Zigzag Silverflexagon.

We have now found all 12 states, 2 square (A and B), 8 hexagonal (C, D, E, F, G, H, J, and K), and 2 rectangular (L and M) and identified the various flexes that allow us to move between them.

The Zigzag Silverflexagon may seem quite complex but it is by far the simplest of the silverflexagons explored in this book.

The Woven Flexatube

The Woven Flexatube

The Woven Flexatube is an extended flexagon obtained by filling the indentations of the Zigzag Silverflexagon with extra segments. For the sake of complete clarity these extra segments have been identified using lower case letters rather than numbers.

The Woven Flexatube forms the basis of Robert E Neale's famous Sheep and Goats puzzle and for this reason is sometimes known as the Sheep and Goats Flexagon. The puzzle involves separating sheep (represented by white triangles) from goats (represented by black ones) - or possibly vice versa. When the puzzle begins both square faces of the flexagon are a mixture of white and black triangles. When it is solved one face is completely white and one completely black. Details of this puzzle treatment have been published in British Origami magazine and elsewhere.

The addition of the eight extra segments to the Zigzag Silverflexagon has the effect of restricting the ways in which the Woven Flexatube can be flexed and of lessening the number of faces that can be found. The Woven Flexatube cannot be flexed by means of rolling flexes, tube flexes, swivel flexes or tuck flexes. It has two square states, no hexagonal states and no oblong ones. Despite this the Woven Flexatube is a complex and fascinating flexagon to explore and map.

The paucity of flat states focuses attention on the multiplicity of routes between them. and on the intermediate forms (square section tubes and two-pocket parallelograms) that are the half-way points along these routes.

Making the Woven Flexatube

A template for this flexagon can be found on page 129.

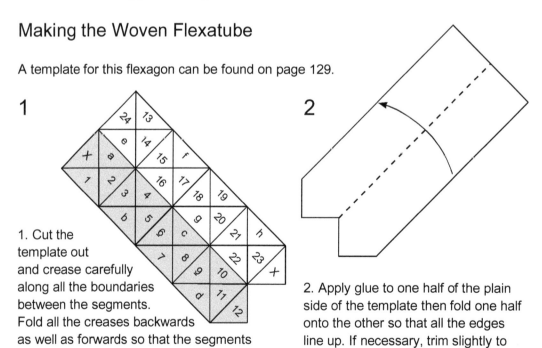

1. Cut the template out and crease carefully along all the boundaries between the segments. Fold all the creases backwards as well as forwards so that the segments move freely in both directions.

2. Apply glue to one half of the plain side of the template then fold one half onto the other so that all the edges line up. If necessary, trim slightly to neaten the edges.

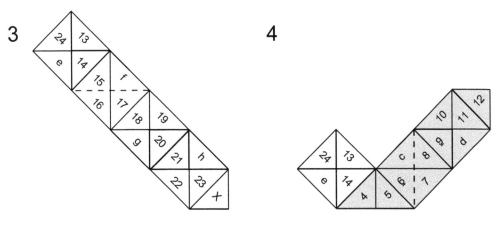

3. Fold 16/17 onto 15/f.

4. Fold 8/7 onto c/6

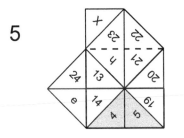

5. Fold 23/22 onto h/21 so that segment x goes underneath 13.

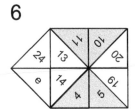

6. This is the result. Turn over sideways.

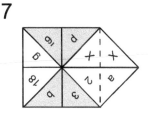

7. Glue x onto x.

8. This is state A of the Woven Flexatube. Turn over sideways.

9. Compare this picture with picture 7 on page 13.

David Mitchell / Silverflexagons and the Flexatube

Diagonal Tube flexes

A diagonal tube flex is made in a similar way to a normal tube flex except that it begins and ends in a diagonal direction. (You can flex between states A and B of the Zigzag Silverflexagon using this kind of flex, though it is far easier in that case to use a double flip flex or normal tube flex instead.)

10
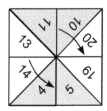
10. Fold 10 onto 20 and 14 onto 4.

11
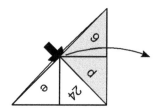
11. Open up the layers at the focus and fold flap 9/d across to the right.

12
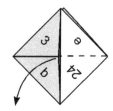
12. Open flap 3/b as shown.

13
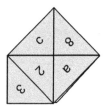
13. Turn over sideways.

14
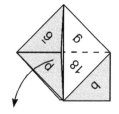
14. Open flap 9/d as shown.

15
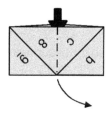
15. Separate the layers to form a square section tube.

16
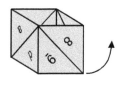
16. Squash flat in the alternate direction.

17
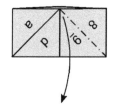
17. Fold the front layers downwards.

18
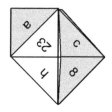
18. Turn over sideways.

19

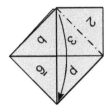

19. Fold b/3 onto 9/d.

20

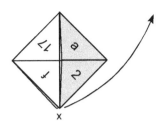

20. Separate the layers at point x and fold the top layers diagonally upward to the right.

21

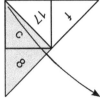

21. Fold the top layers diagonally downwards to the right.

22

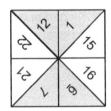

22. This is state B.

Making a diagonal tube flex from any face of either square state in any direction will take you to the other square state.

Diagonal tube flex map of the Woven Flexatube

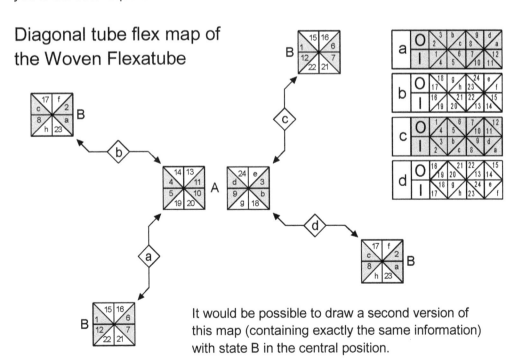

It would be possible to draw a second version of this map (containing exactly the same information) with state B in the central position.

The four possible tube forms of the Woven Flexatube are identified on this map by the letters a, b, c and d placed inside the diagonal tube flex symbols. The four boxes to the top right of the map show which surfaces form the outside (O) and inside (I) of each of these four tubes. The background of the boxes has been shaded to show the pattern produced on the outside and inside of these tubes when the strip is cut from differentiated paper.

You will notice that the outside of tube a is the same as the inside of tube c and vice versa. This allows us to design a puzzle treatment of the flexagon in which the object is to turn a tube inside out, a challenge very similar to that of the Flexatube which is explained elsewhere in this book. If you investigate this possibility you will find that there are two quite separate ways in which such a challenge can be solved. The obvious solution is to flex straight through state A or B but it is also possible to avoid going through either state. This second solution can be found as Easy Street in the solutions given for the Flexatube (see pages 98 and 99).

You will also notice that the visible surface of tubes a and c is entirely shaded and that of tubes d and b entirely white. This allows us to design a further puzzle treatment in which the object is to turn a tube of entirely one colour or pattern into a tube of entirely another. Here again the solution is to flex through square states A or B but in this case the axis of the flex must be altered at right angles when the square state is reached.

Taking the Scenic Route

Somewhat remarkably it is possible to flex between states A and B without making a tube flex. Here's how to do it.

23

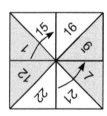

23. Just for a change this is state B. Fold 1 onto 15 and 21 onto 7.

24

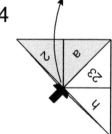

24. Open the focus and fold 2/a upwards.

25

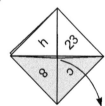

24. Open flap 8/c as shown.

26

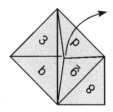

26. Open flap 9/d as shown.

27

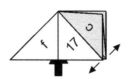

27. Open to form a pyramid.

28

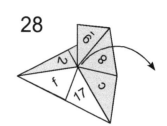

28. Open flap c/8.

29

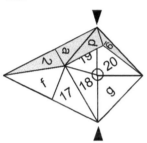

29. Squeeze the sides together so that the point marked with a circle rises upwards.

30

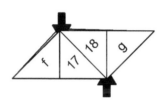

30. This parallelogram form has pockets to top and bottom. Rotate through 180 degrees.

31

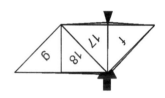

31. Open the pocket at the bottom and push down on the front top edge between 17 and f.

32

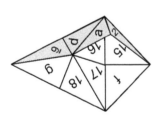

32. Fold f onto 17 and 15 onto 16.

33

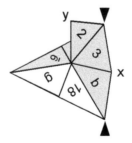

33. Squeeze flat sideways so that x and y come together.

34

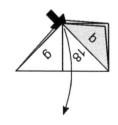

34. Fold flap 18/g downwards.

David Mitchell / Silverflexagons and the Flexatube

35

35. Fold a/2 onto 8/c.

36

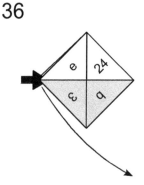

36. Separate the layers at the left side and fold the front left corner diagonally downwards.

37

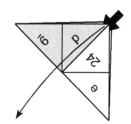

37. Separate the layers at the top right corner and fold the front layer diagonally downwards.

38

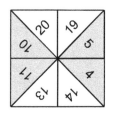

38. This is state A.

There are eight of these scenic routes altogether, each of which goes through a different two-pocket parallelogram. You may like to investigate the relationship between these eight parallelograms, the four tubes and the two square states in more detail for yourself.

Scenic route map of the Woven Flexatube

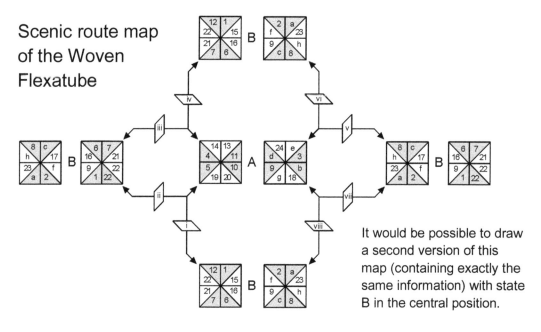

It would be possible to draw a second version of this map (containing exactly the same information) with state B in the central position.

Other Related Silverflexagons

By beginning with the strip used to make the Woven Flexatube (strip A) and removing one or more segments it is possible to arrive at twelve other strips, all of which can be turned into silverflexagons. One of these, strip L, is the strip for the Zigzag Silverflexagon. You will notice that strips K, L and M will form pure flexagons while all the others will form extended flexagons. I have not carried out an exhaustive exploration of the properties of these forms. It would seem logical that their properties will be partly those of the Zigzag Silverflexagon and partly those of the Woven Flexatube. However, given the uncertain nature of flexagons, there may well be some surprises in store for anyone who cares to explore a little.

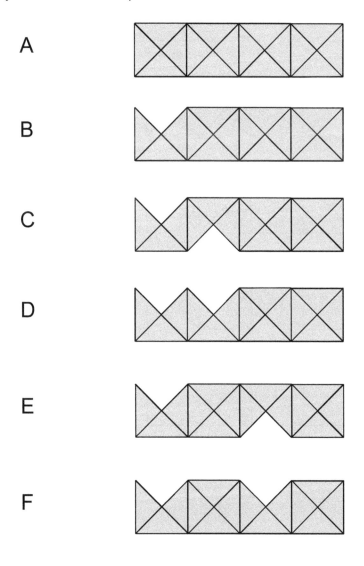

G
H
I
J
K
L
M

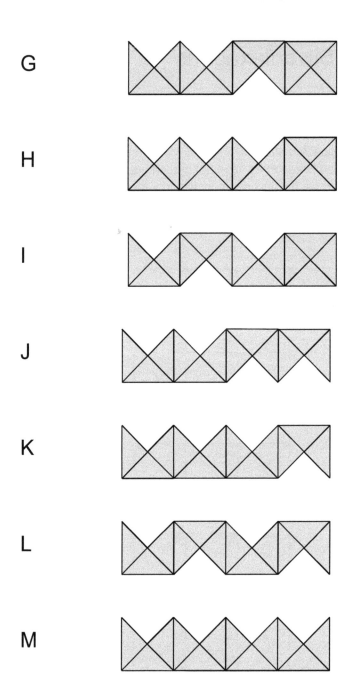

Slit-Square Silverflexagons

The First Slit-Square Silverflexagon

As the name implies, the First Slit-Square Silverflexagon can be made by first folding, then slitting, a square of paper in the way shown below. The result is a continuous strip of sixteen silver triangles. This strip can then be twisted and the ends re-joined to form the flexagon. Diagrams for making the flexagon in this way can be found on pages 54 to 56.

This method is recommended if you want an easy way to make the First Slit-Square Flexagon to use as a flexible toy. For study purposes, however, it is better to make one using the pre-numbered template.

The Slit-square Silverflexagon is a pure flexagon, which is to say that each of the 32 segments stretches across the entire width of the strip from edge to edge.

The increase in complexity between the Zigzag Silverflexagon and the First Slit-Square Silverflexagon can be judged by comparing the number of states that each possesses. As already seen, the Zigzag Silverflexagon has two square, eight hexagonal and two oblong states. The First Slit-Square Silverflexagon has seventeen square states, sixteen hexagonal states and four oblong states. The reason why the First Slit-Square Silverflexagon has an odd number of square states is that it has just one home position, whereas the Zigzag Silverflexagon has two equally symmetrical home positions.

The first Slit-Square Silverflexagon is a twisted flexagon. There are also two non-twisted variants which can be made from a slit square of the same pattern in which however the slit does not extend to the edge of the square, thus leaving the segments as a continuous band. The properties of these Non-Twisted Slit Square Flexagons are explored later in this chapter.

Making the First Slit-Square Silverflexagon

A template for the First Slit-Square Silverflexagon in which each segment is individually numbered can be found on page 130. This template is in two halves which must be folded up then joined together to create the finished flexagon.

1

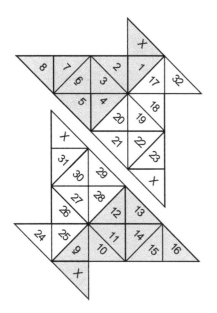

2

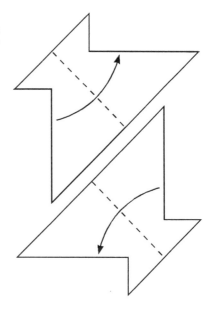

1. Cut the template out and crease carefully along all the boundaries between the segments. Fold all the creases backwards as well as forwards so that the segments move freely in both directions.

2. Apply glue to one half of the plain side of each section then fold one half onto the other so that all the edges line up. Do not glue the two sections together! If necessary trim to neaten the edges.

3

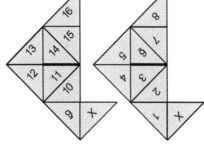

3. Cut carefully along the creases marked with thick black lines.

4

4. Fold 10 onto 9 and 2 onto 1.

5

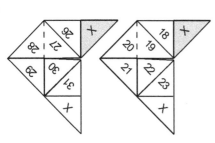

5. Fold 28 onto 27 and 20 onto 19, allowing the paper below the fold to flip to the left behind.

David Mitchell / Silverflexagons and the Flexatube

6.

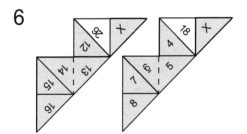

6. Fold 14 onto 13 and 6 onto 5.

7.

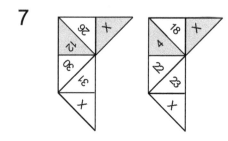

7. Rotate the right hand section through 180 degrees.

8.

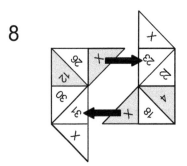

8. Slide the two parts together so that the shaded segments marked with x overlay segments 23 and 31.

9.

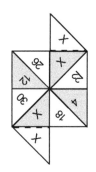

9. Glue x onto x twice.

10.

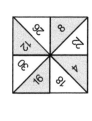

10. This is state A of the First Slit-Square Silverflexagon.

Flexing the First Slit-Square Silverflexagon

Here are some maps showing how to flex the First Slit-Square Flexagon using flexes we have already met.

Double Flip Flexes

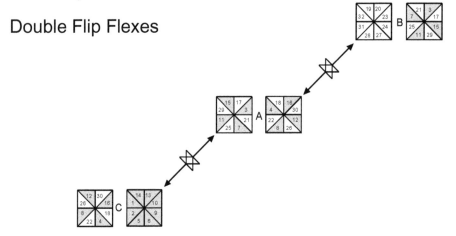

David Mitchell / Silverflexagons and the Flexatube

Rolling Flexes and Tube Flexes

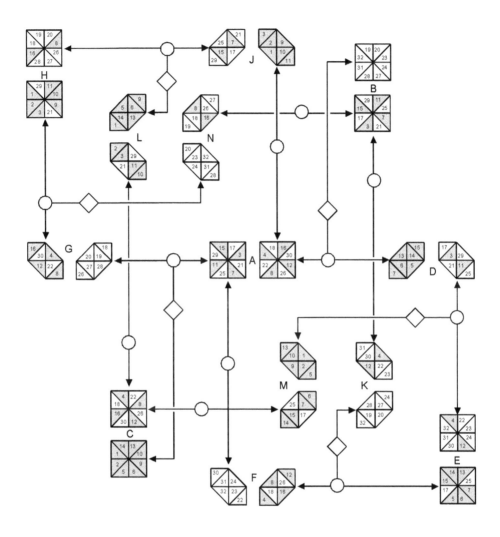

Alternative routes

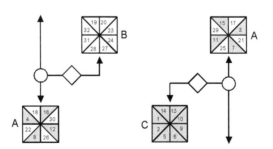

Double Swivel Flexes

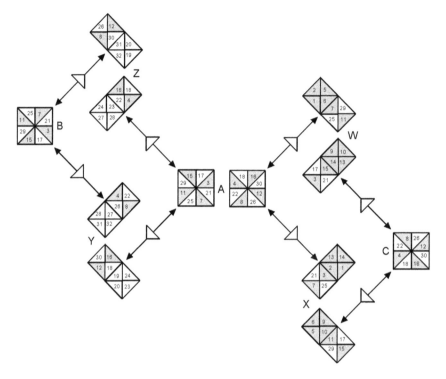

Single Tuck Flexes

There are two independent cycles of single tuck flexes to be found.

Cycle 1

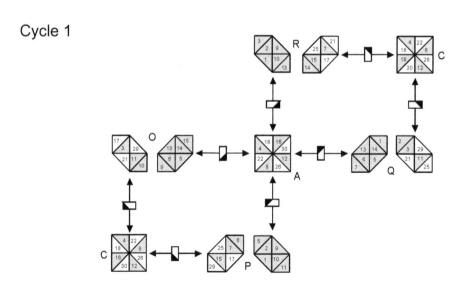

Cycle 2

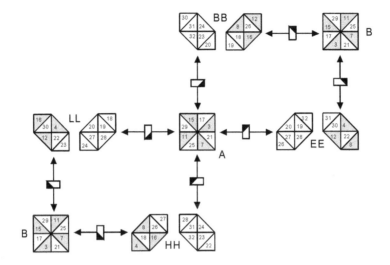

It is interesting to compare these maps with the single tuck flex map for the Zig-Zag Silverflexagon on page 24.

Double Tuck Flexes

These three maps do not link to each other, but the hexagonal states reached from states A, B and C using double tuck flexes can be connected to each other using open tuck flexes.

Map 1

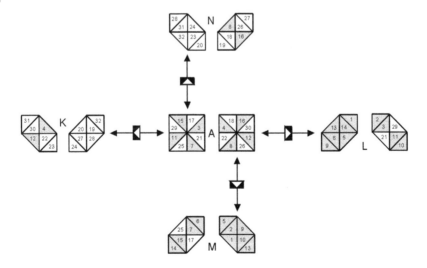

Map 2

Map 3

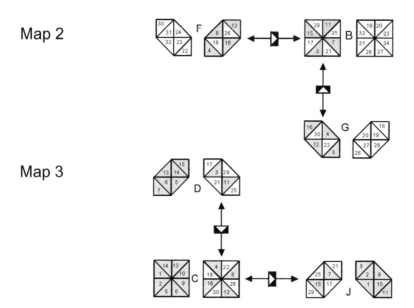

Open Tuck Flexes

There are four unconnected cycles of open tuck flexes. Every hexagonal state appears in one of these cycles.

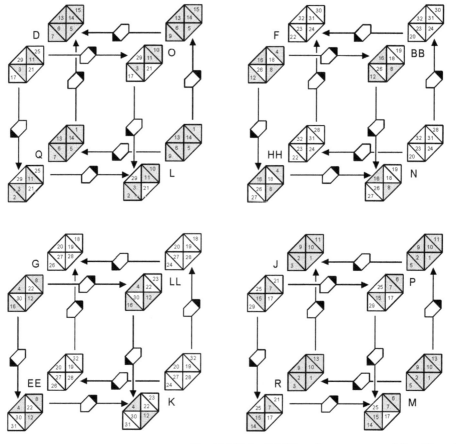

Reverse Tube Flexes

If you study the rolling flex and tube flex map on page 41 you will see that if you were to begin with C and flex to G you would begin the flex by folding the right hand half of the flexagon across to the left but end by folding the left hand half of the flexagon across to the right. We can call this type of flex a reverse tube flex.

States O, P, Q and R (see the Cycle 1 single tuck flex map on page 42) provide access to eight more square faces, four by means of rolling flexes and four by means of reverse tube flexes. These faces are themselves linked in four cycles that include faces BB, EE, HH and LL. Because of this it would be possible to draw a second version of the map below based on the Cycle 2 single tuck flex map.

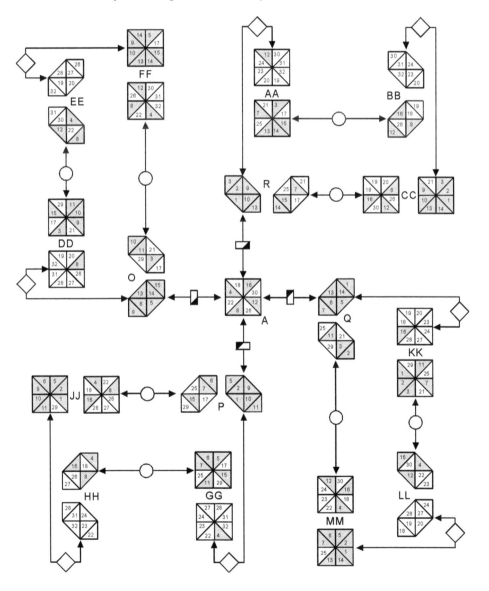

Swivel / Flip Flexes

As their name suggests, swivel / flip flexes are a type of diagonal flex that are half a swivel flex and half a flip flex. Here's how to make one.

28.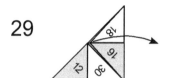

28. Begin with State A. Fold 11 onto 25 and 17 onto 3.

29. Swivel 18/16 to the right.

30.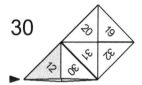

30. Push the lower left corner inside out between the layers.

31.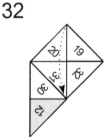

31. Separate the layers at the left edge and flip the internal point downwards.

32. Swivel 20/19 downwards behind the other layers.

33.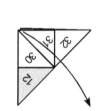

33. Fold the front layers downwards to the right.

34.

34. This is face U.

Swivel / Flip Flex map of the First Slit-Square Silverflexagon

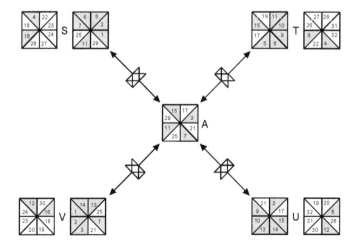

46   David Mitchell / Silverflexagons and the Flexatube

Non-Twisted First Slit-Square Silverflexagons

There are two versions of the Non-Twisted First Slit-Square Silverflexagon. In Version 1 the two halves of the flexagon are rotated in the same direction. In Version 2 they are rotated in opposite directions.

In some states both of these flexagons exhibit a degree of instability and so you will need to be careful to maintain the integrity of the states as you flex between them. In some of the less stable states you will find that the focus can be pulled apart without first bringing edges or corners together.

Version 1 has seven true states altogether, five square and two oblong. Version 2 has thirteen true states, five square and eight hexagonal. Version 1 can be flexed using double flip, tube, double swivel and swivel/flip flexes. Version 2 can be flexed using double flip, rolling, tube and reverse tube flexes. In addition Version 1 can be flexed using pop-up swivel flexes. This is not a true flex (since it begins by pulling the focus of a face apart without first bringing any corners or edges together) but it is nevertheless interesting to explore, analyse and map.

The instructions show you how to make these flexagons from the template on page 130. You can also make them directly from a slit square. See page 56.

Making Version 1 of the Non-Twisted First Slit-Square Silverflexagon

Begin by following instructions 1 to 3 on page 39 then arrange the two halves in the way shown in picture 4 below.

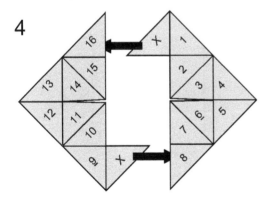 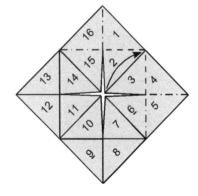

4. Apply glue to both segments marked with an x then glue the two halves of the flexagon together so that the segments marked with x go underneath segments 8 and 16 respectively.

5. The strip is complete. Rotate the square formed by segments 2 and 3 through 180 degrees in the direction indicated by the arrow.

6 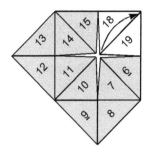 7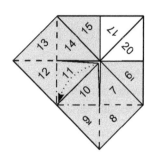

6. Continue rotating this square in the same direction another 180 degrees until all the segments lie flat.

7. Repeat steps 2 and 3 on the other half of the strip making sure you rotate square 11/10 in the same direction as you rotated square 2/3.

8 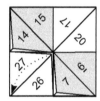 9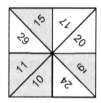

8. Continue rotating this square in the same direction through another 180 degrees until all the segments lie flat.

9. Version 1 of the Non-Twisted Slit-Square Silverflexagon is finished.

Making Version 2 of the Non-Twisted First Slit-Square Silverflexagon

Begin by following instructions 1 to 3 on page 39 then instructions 4 to 6 on page 47 and above.

7 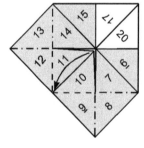 8 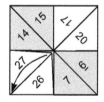 9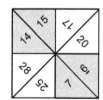

7. Repeat the rotation on the other half of the strip but make sure you rotate square 11/10 in the opposite direction this time.

8. Continue rotating this square in the same direction through another 180 degrees until all the segments lie flat.

9. Version 2 of the Non-Twisted Slit-Square Silverflexagon is finished.

Flexing Version 1 of the Non-Twisted First Slit-Square Silverflexagon

Double Flip Flexes

Tube Flexes

Swivel / Flip Flexes

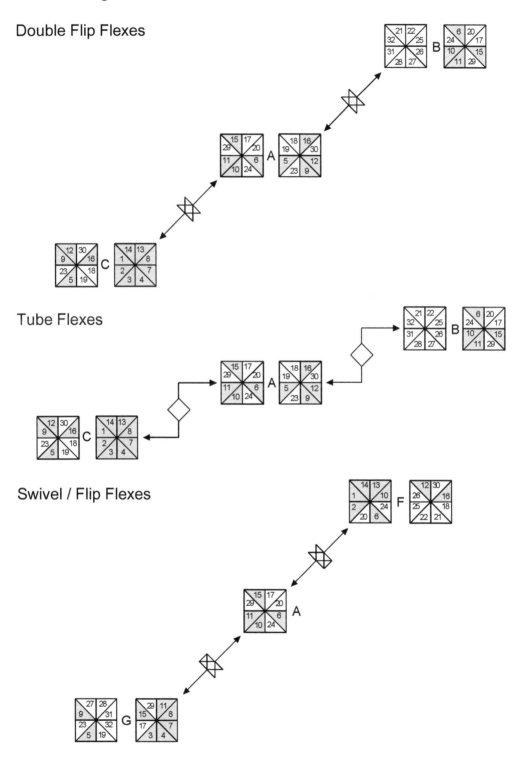

Double Swivel Flexes

In the double swivel flexes we have met before one of the flaps is swivelled forwards and the other backwards. In the case of Version 1 of the Non-Twisted First Slit-Square Silverflexagon, however, both flaps must be swivelled inn the same direction.

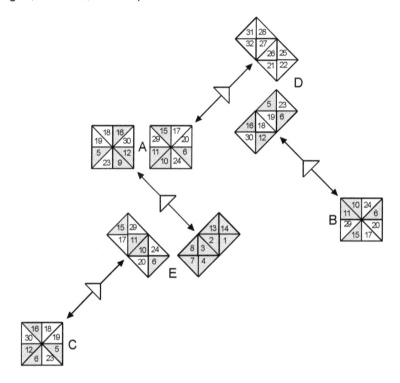

Pop-up swivel Flexes

Pop-up swivel flexes are made like this:

10

11

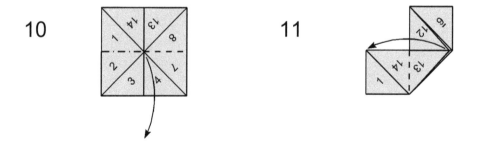

10. Begin with state C of version 1. Pull the focus apart to separate the front layers (13, 14, 1, 2, 3 and 4) from the back layers (7 and 8) and fold 3 and 4 downwards. Segment 7 remains still.

11. Swivel 12/13 to the right.

12 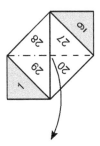   13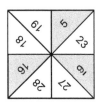

12. Pull the focus apart to separate the front layers (9, 27, 28, 29 and 1) from the back layers (20) and fold 1 and 29 downwards. Segment 20 remains still.

13. This is the result.

When you try this you will find that steps 10 and 12 involve forming the flexagon into a tube-like form then flattening it in the alternate position.

It is also possible to make pop-up swivel flexes in a sideways direction from face C. In fact, as the map overleaf shows, the pop-up swivel flex map of Version 1 of the Non-Twisted Slit-Square Silverflexagon can be drawn as a square grid. Each column and row of the grid is a sequence of flexes that cycles back to the starting point after four moves.

For the sake of clarity only one face of each state has been shown.

The top row and the bottom row of faces are identical as are the left and right columns. Face C occurs in every corner and face B is in the centre. The map could, of course, equally have been drawn the other way round, with C in the centre and B at the corners.

You will notice that states F and G (see swivel flex map) also occur here. All the other states, those not identified by letters, can only be accessed by pop-up swivel flexes.

You may like to investigate for yourself what happens if you try to make pop-up swivel flexes from state A or using Version 2.

Pop-up Swivel Flexes

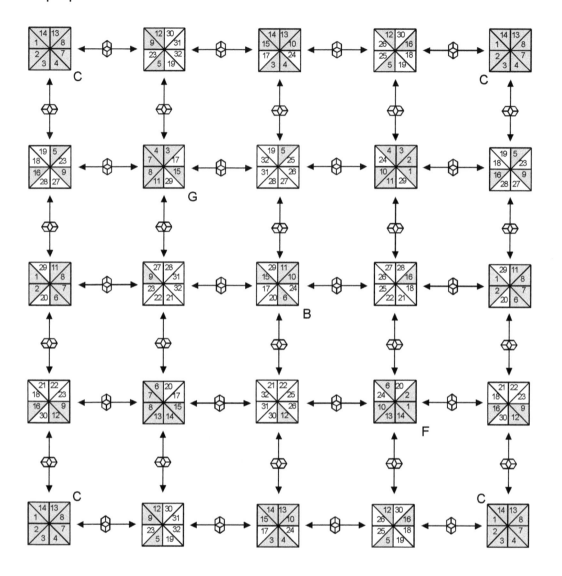

Flexing Version 2 of the Non-Twisted First Slit-Square Silverflexagon

Rolling Flexes, Tube Flexes and Reverse Tube Flexes

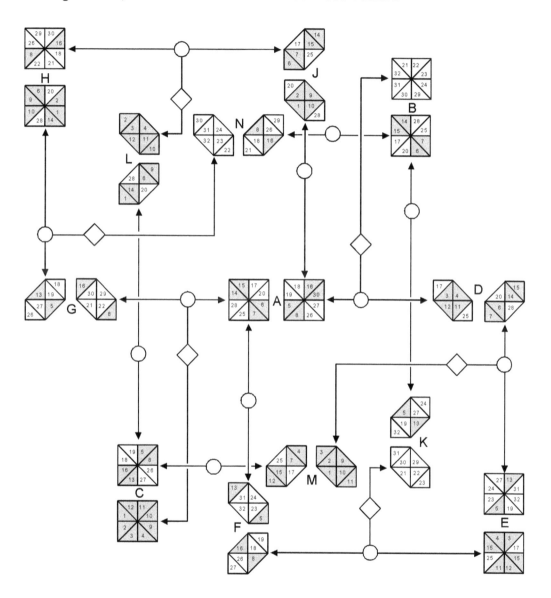

It is interesting to compare this map with the map of similar flexes for the First Slit-Square Silverflexagon on page 41. They are similar, though not identical.

It is, of course, also possible to move between states A and B, and faces A and C, by means of double flip flexes.

David Mitchell / Silverflexagons and the Flexatube

Making the First Slit-Square Silverflexagon from a single square

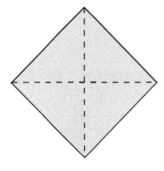

1. Crease in both diagonals like this.

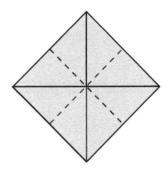

2. Fold in half both ways edge to edge as well.

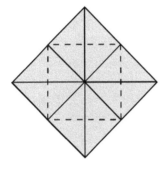

3. Fold all the corners into the centre to create these four new creases, then unfold,

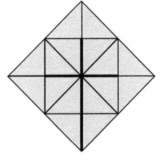

4. Check you have made all eight creases. Make sure they fold easily in both directions, backwards as well as forwards. Cut a cross shaped slit along the creases marked with thick black lines.

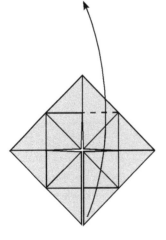

5. Fold the right hand half of the paper upwards as shown.

David Mitchell / Silverflexagons and the Flexatube

6

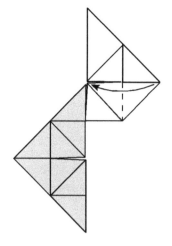

6. Fold the right point inwards as shown allowing the paper above the fold to swivel into the position shown in picture 7.

7

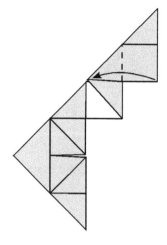

7. Fold the right edge inwards like this.

8

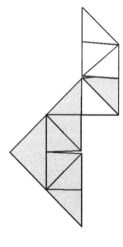

8. Turn over sideways.

9

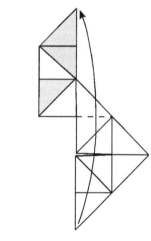

9. Fold the bottom point upwards like this.

10

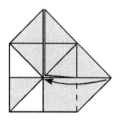

10. Fold the right point inwards as shown allowing the paper above the fold to swivel into the position shown in picture 11.

11

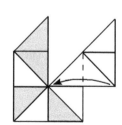

11. Fold the right edge inwards like this.

David Mitchell / Silverflexagons and the Flexatube

12

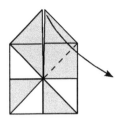

12. Fold the top right point diagonally outwards as shown.

13

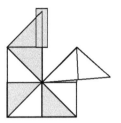

13. Attach a piece of clear sticky tape to the edge of the top segment like this.

14

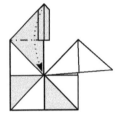

14. Trim the sticky tape to shape then fold the top segment downwards behind the other layers.

15

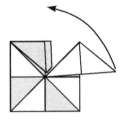

15. Return the segments you folded outwards in step 12 to their original position ensuring that they go behind the sticky tape.

16

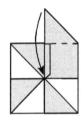

16. Fold the top segment downwards and press it onto the sticky tape.

17

17. The First Slit-Square Silverflexagon is finished.

Making the Non–Twisted First Slit-Square Silverflexagons from a single square

Both non-twisted versions of the First Slit-Square Silverflexagon can be made by slitting just the central part of the cross marked with thick black lines in step 4, so that the slit does not extend to the edge of the square, then following instructions 5 through 8 (in both cases) on pages 47 and 48.

Other Slit-Square Silverflexagons

The patterns pictured below show how the pattern on which the slit-square silverflexagons we have already explored were based (the pattern on the left below) can be extended to produce a series of ever more complex strips which could also be folded up to form silverflexagons. Non-twisted variants could, of course, be made from strips of an identical pattern in which the slits were confined just to the central area.

I have not explored these possibilities to any great extent.

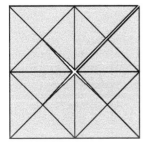 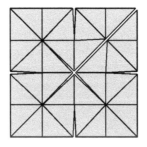 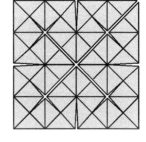

Pattern for the Second Slit-Square Silverflexagon.

Pattern for the Third Slit-Square Silverflexagon.

Generalising the design to other polygons

The illustrations below show how the slit-square design can be generalised to apply to other polygons.

One of these (pictured below) will be a bronzeflexagon (a flexagon made from a strip divided into segments which are all identically sized right angle triangles with sides in the proportion of 1:2:sqrt3). The rolling flex and tube flex map for this flexagon (and any other first level slit-diamond flexagon) will be essentially similar to the map on page 41. You may like to work out for yourself which of the other flexes will work and which will not and why.

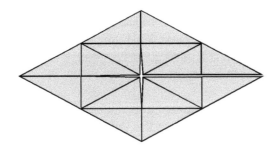 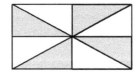

David Mitchell / Silverflexagons and the Flexatube

The slit-square design will also generalise to enable us to construct slit-polygon flexagons with any given number of sides like this.

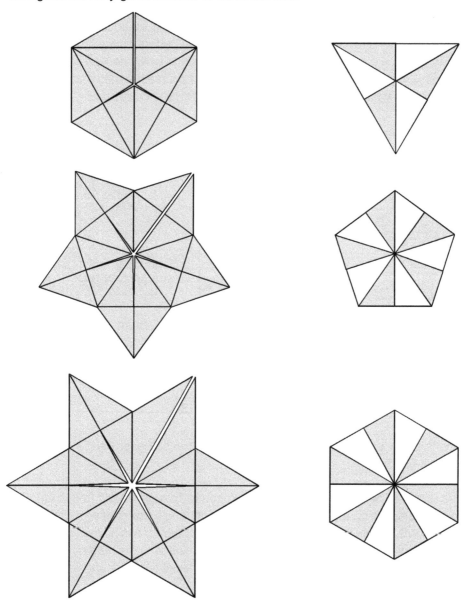

The first and third of these are also bronzeflexagons. The Slit-Hexagon Bronzeflexagon is particularly interesting, but an analysis of its many pyramidal forms is beyond the scope of this present enquiry. The second form is a pentagonal flexagon. It is interesting for this reason (but, alas, only interesting for this reason).

All slit-polygon flexagons with more than four sides appear to be inherently unstable in certain states.. The more sides the polygon has, the worse this tendency towards chaos appears to become.

The Labyrinth Silverflexagon

The Labyrinth Silverflexagon

The Labyrinth Silverflexagon is much more complex than the silverflexagons we have already met. The increase in complexity from the First Slit-Square Silverflexagon can be judged by comparing the number of states of each shape that each flexagon possesses. The First Slit-Square Silverflexagon has seventeen square states, sixteen hexagonal states and four oblong ones. The Labyrinth Silverflexagon also has four oblong states but there the similarities end. A full count reveals that the Labyrinth Silverflexagon also possesses thirty square states, fifty-six hexagonal states, one hundred and twenty-two pentagonal states and sixteen quadrilateral ones. Welcome to the labyrinth!

Perhaps the best way to understand the structure of the Labyrinth Silverflexagon is to think of it as a mine that has been driven into a hillside at several different levels. The two main levels are not only labyrinthine in their own right but are also connected to each other inside the hill by routes and intermediate states which we can think of as shafts and ladderways, some of which link the two levels directly, but others of which lead to and from a number of sub-levels and transfer-platforms hung between them like mezzanine floors.

In order to make this structure clear on the maps those states that belong to each of the main levels are identified by labels beginning with 1 or 2 (so 1A and 2B etc) States that form connections between the two main levels, and so belong exclusively to neither, are identified with labels in the form A, B or AA, BB etc. States that belong to the sub-levels are identified by labels in the form S1, S2 etc where S is the identifying letter for the connecting state.

Making the Labyrinth Silverflexagon

A template for the Labyrinth Silverflexagon in which each segment is individually numbered can be found on page 131.

1

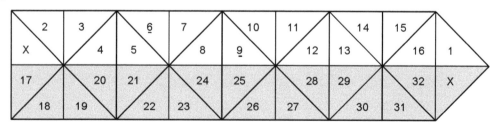

1. Cut the template out and crease carefully along all the boundaries between the segments. Fold all the creases backwards as well as forwards so that the segments move freely in both directions.

2

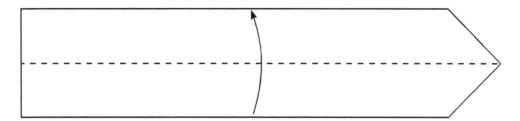

2. Apply glue to one half of the plain side of the template then fold this half onto the other half so that the edges line up. If necessary trim slightly to neaten the edges.

3

3. Fold 29 onto 30.

4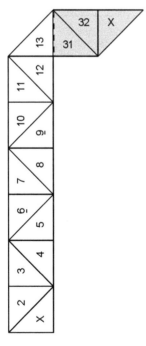

4. Fold 13 onto 31.

5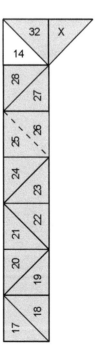

5. Fold 25 onto 26.

David Mitchell / Silverflexagons and the Flexatube

6

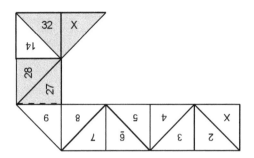

6. Fold 9 onto 27.

7

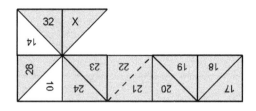

7. Fold 21 onto 22.

8

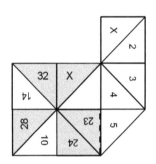

8. Fold 5 onto 23 making sure that 3 goes underneath X.

9

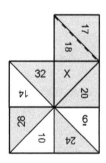

9. Fold 17 onto 18.

10

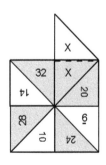

10. Glue X onto X.

11

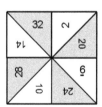

11. This is state 1A (state A of level 1) of the Labyrinth Silverflexagon.

Flexing the Labyrinth Silverflexagon

The Labyrinth Silverflexagon has four home positions, 1A, 1B, 2A and 2B, two on each of the main levels. As the first map shows, you can flex between states 1A and 1B, and between states 2A and 2B, using double flip flexes.

Inside Out Flexes

You can move also between the corresponding home positions (1A to 2A and 1B to 2B) by using inside out flexes. Here's how to do it.

12

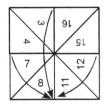

12. Begin with state 1A. Fold 7 onto 8 and 12 onto 11, then collapse symmetrically.

13

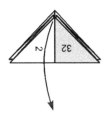

13. Fold the front layers downwards.

14

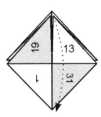

14. Flip the front right hand point downwards between the layers.

15

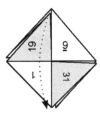

15. Repeat this manoeuvre on the front left hand point.

16

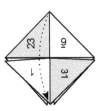

16. Fold the remaining layers downwards behind.

17

17. Reveal the new face by pulling the front layers downwards.

18

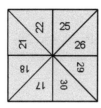

18. This is state 2A.

Double Flip Flex and Inside Out Flex map of the Labyrinth Silverflexagon

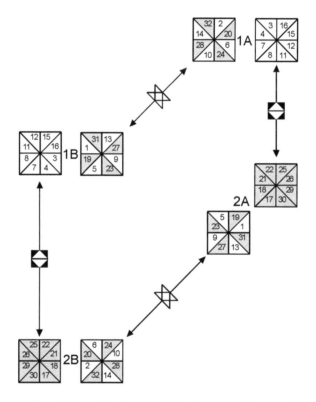

If you work around all four states, turning each state over sideways before you perform the next flex in the way the map shows you should, you will find that the flexagon will not line up with the map when you reach the beginning again. Odd.

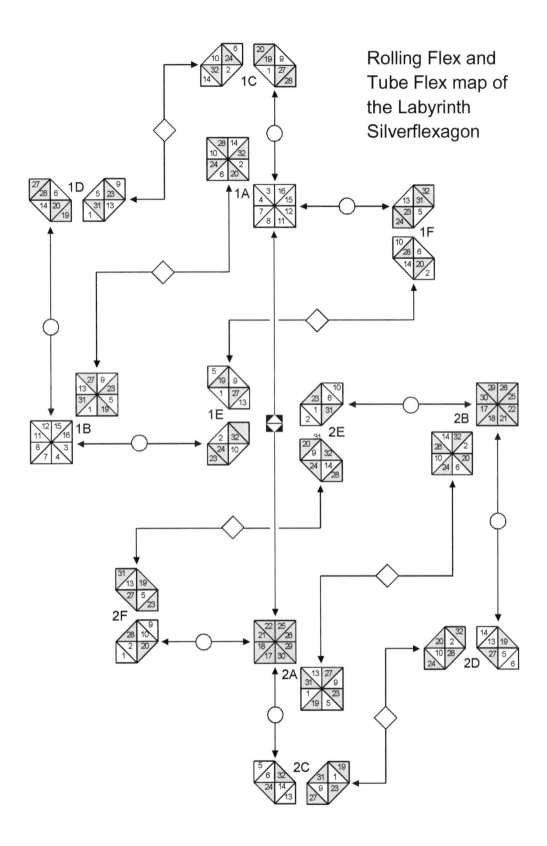

Rolling Flex and Tube Flex map of the Labyrinth Silverflexagon

David Mitchell / Silverflexagons and the Flexatube

Single Tuck Flex Map of the Labyrinth Silverflexagon

Level 1

Level 2

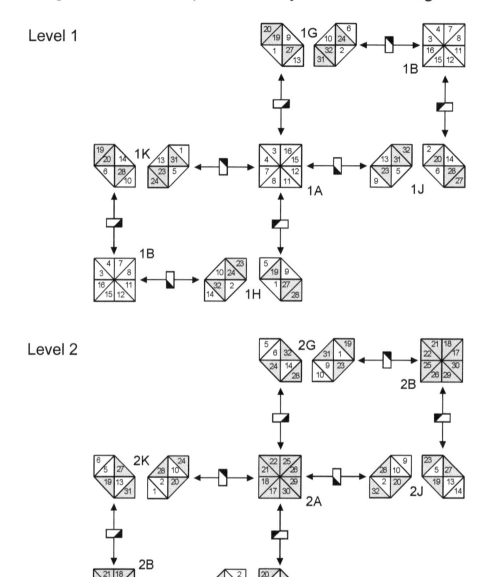

Double Tuck Flexes

As you would expect, you can also flex the Labyrinth Silverflexagon using double tuck flexes and move directly from 1A to 1D and 1F and from 1B to 1C and 1E in this way.

Similarly, on level 2, you can use double tuck flexes to move directly from 2A to 2D and 2F and from 2B to 2C and 2E.

Double Swivel Flex Map of the Labyrinth Silverflexagon

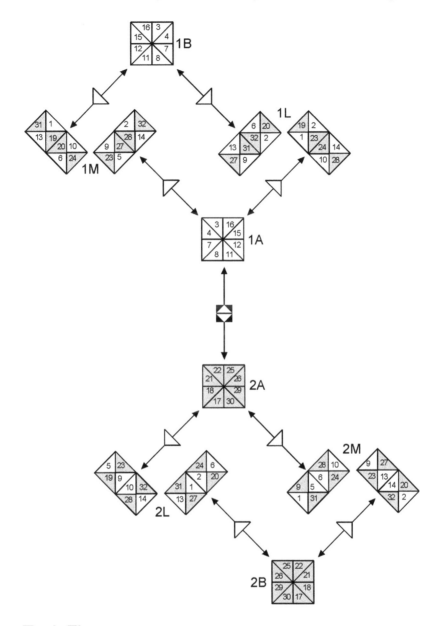

Open Tuck Flexes

Open tuck flexes occur in cycles of four related states. The Labyrinth Silverflexagon has twelve such cycles. Four of these link hexagonal states and the remaining eight link pentagonal states. Two of the four hexagonal states belong to each of the main sub-levels. However, in each of the pentagonal cycles two of the states belong to the main sub-levels, one to each, while the remaining states are intermediate points on the routes between them.

Open tuck flexes are made in exactly the same way irrespective of whether the corner you are tucking belongs to a hexagonal or a pentagonal state.

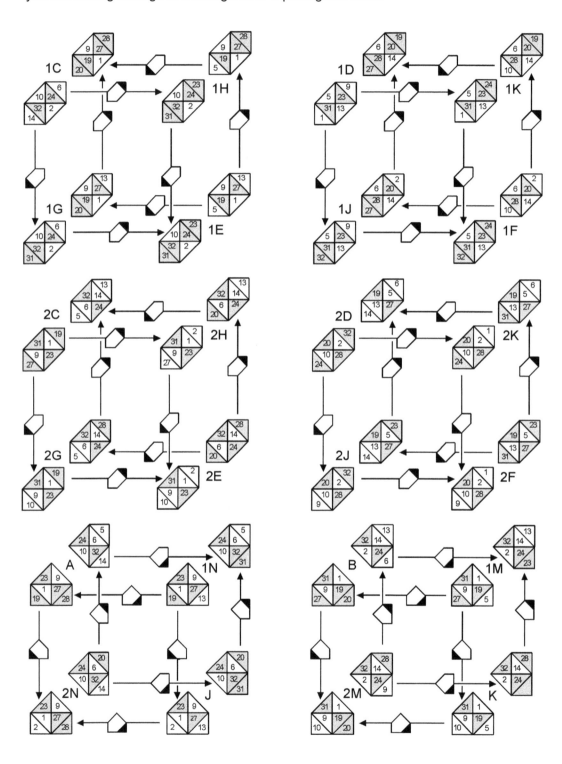

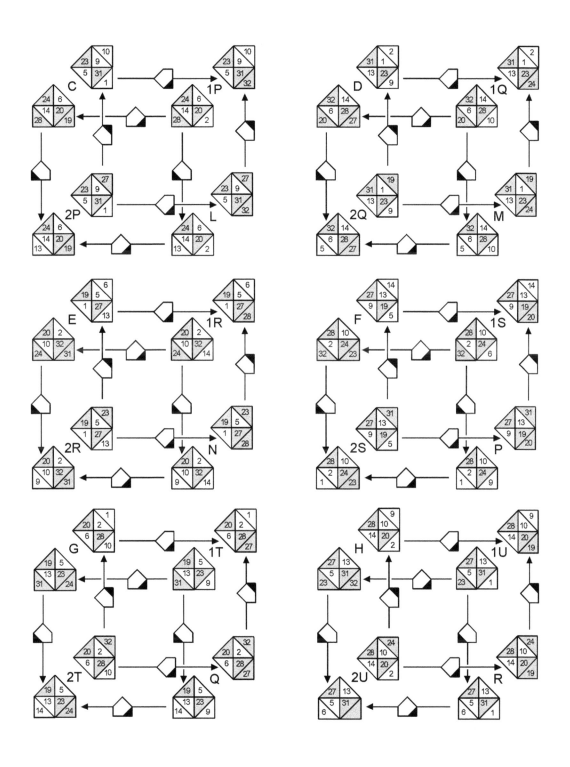

David Mitchell / Silverflexagons and the Flexatube

Open Squash Flexes

Open squash flexes change hexagonal states into pentagonal states and vice versa. They are extremely simple to perform.

19

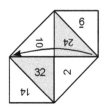

19. Begin with state 1C. Fold 24 onto 10.

20

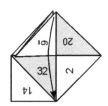

20. Fold 6/20 onto 32/2.

21

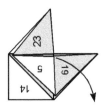

21. Open out the front layers as shown.

22

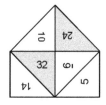

22. This is state A.

Like open tuck flexes, open squash flexes occur in cycles. The Labyrinth Silverflexagon has four such cycles. The two hexagonal states in each cycle belong to the two main sub-levels, while the pentagonal states are intermediate states on the routes between them.

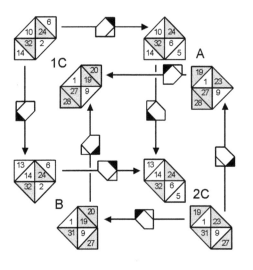

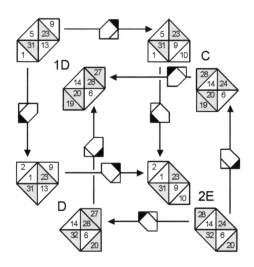

David Mitchell / Silverflexagons and the Flexatube

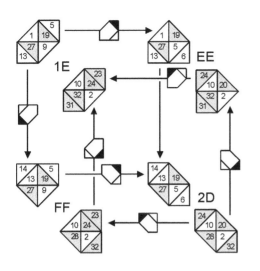
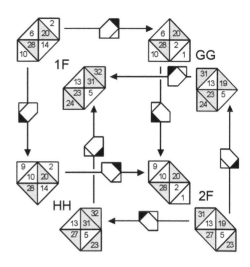

Links between the cycles

These sixteen cycles of hexagonal and pentagonal states are linked in many different ways. Some of these links are obvious (because some states occur in several different cycles) but there are others which cannot be easily deduced from the maps.

Open squash flexes can, for instance, be used to link state 1G to state 1N and state 1S in the way shown here.

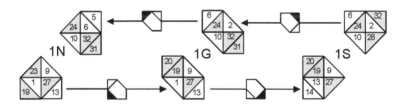

Similarly, 1H links to 1R and 1M, 1J links to 1P and 1T and 1K links to 1Q and 1U. A corresponding series of similar links exists on level 2.

Applying an open squash flex to state J produces a surprising result.

23

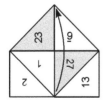

24

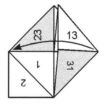

23. Begin with state J. Fold 27 onto 9.

24. Fold 13/31 onto 23/1.

David Mitchell / Silverflexagons and the Flexatube

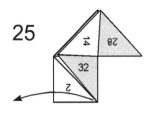 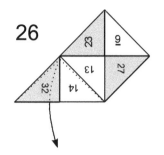 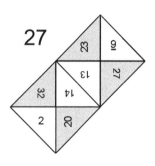

25. Fold 2 across to the left and allow 28 to fold downwards.

26. This is the result. Swivel the hidden flap into view.

27. This is state 2M.

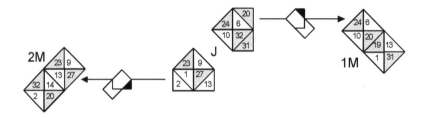

To move from 2M or 1M to J simply follow the instructions in reverse. Similar links exist for K, L, M, N, P, Q and R. You may like to investigate and map them for yourself.

Accessing the primary sub-levels

There are sixteen primary sub-levels, each of which can be accessed from one of the four home states by means of a complex double tuck and inside swivel flex.

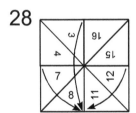 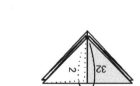 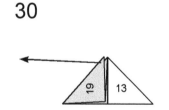

28. Begin with state 1A. Fold 7 onto 8 and 12 onto 11, then collapse symmetrically.

29. Fold flap 2/32 downwards in front then up inside the remaining layers making sure it goes behind the middle layers on the left hand side (shown by the dotted lines).

30. This is the result. Pull the loose flaps out to the left.

31   **32**   **33**

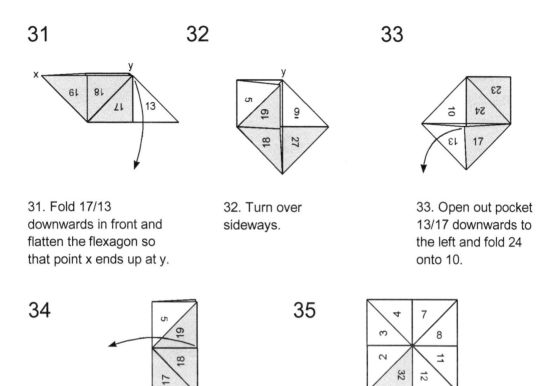

31. Fold 17/13 downwards in front and flatten the flexagon so that point x ends up at y.

32. Turn over sideways.

33. Open out pocket 13/17 downwards to the left and fold 24 onto 10.

34   **35**

34. Fold the front layers to the left.

35. This is state 1X.

To return to the main level simply follow these instructions in reverse.

Map of primary sub-level connections to home state 1A

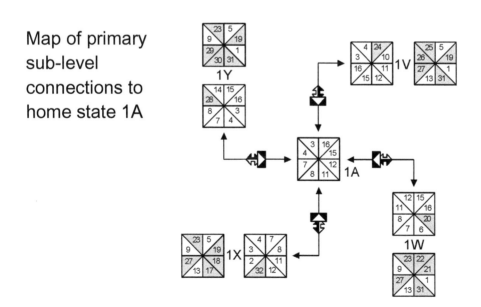

Map of primary sub-level connections to home state 1B

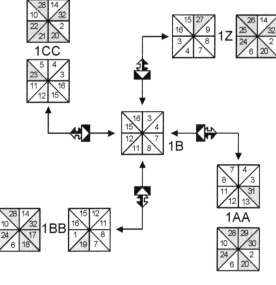

Map of primary sub-level connections to home state 2A

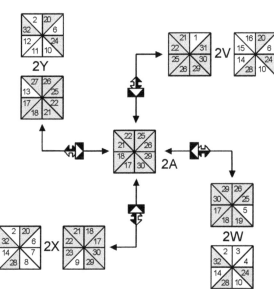

Each of the sixteen square states 1V through 1CC and 2V through 2CC shown on these three maps and the fourth on page 75 gives access to a sub-level consisting of two hexagonal and four pentagonal states. The connections within sub-level 1X are shown on pages 75 and 76. You will note that the states within this sub-level have been identified by the addition of numbers 1 to 6 to the sub-level identifier 1X, thus 1X1 etc.

Space does not permit the inclusion of individual maps for the other primary sub-levels, nor is this necessary, since they are, of course, due to symmetry, structured in the same way, and the identifiers for each state within them can be easily inferred.

Taken together these sixteen primary sub-levels contain 16 square states, 32 hexagonal states and 64 pentagonal ones.

Map of primary sub-level connections to home state 2B

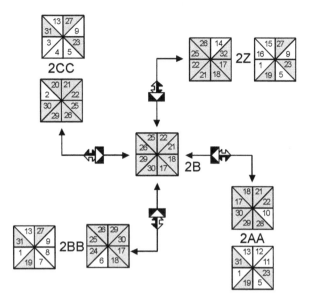

Rolling flex and single tuck flex map of sub-level 1X

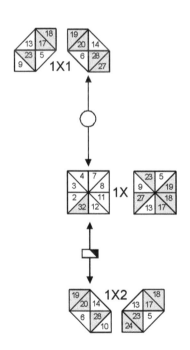

Open single tuck flex map of sub-level 1X

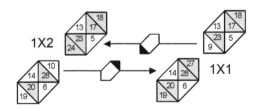

David Mitchell / Silverflexagons and the Flexatube

Open single squash (and associated single tuck) flex maps of sub-level 1X

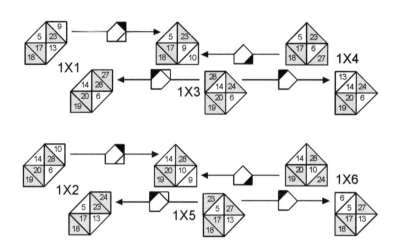

Accessing the secondary sub-levels

There are four secondary sub-levels. As the maps make clear, the square states belonging to each of these sub-levels form intermediate points on routes between the main levels of the flexagon. Like the primary sub-levels these secondary levels are accessed by a complex double tuck and inside swivel flex, the crucial difference being the way in which the loose flap is swivelled inside in steps 37 and 40.

36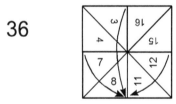

37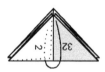

36. Begin with state 1A. Fold 7 onto 8 and 12 onto 11, then collapse symmetrically.

37. Fold flap 2/32 downwards in front then swivel it up inside the remaining layers, making sure it ends up in front of the middle layers on the left hand side.

38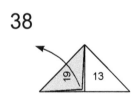

39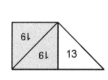

40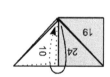

38. This is what the result should look like. Fold 19 upwards to the left.

39. Turn over sideways.

40. Repeat step 37 on flap 10/24.

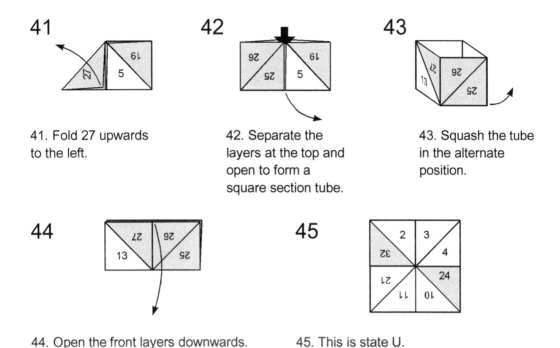

41. Fold 27 upwards to the left.

42. Separate the layers at the top and open to form a square section tube.

43. Squash the tube in the alternate position.

44. Open the front layers downwards.

45. This is state U.

To return to the main levels simply follow these instructions in reverse.

The square states of the four secondary sub-levels have been labelled S through V. A map of secondary sub-level S is given below. You will see that the states within this level have been identified by the addition of the numbers 1 and 2 to the sub-level identifier S.

Space does not permit the inclusion of similar maps for the other three secondary sub-levels, nor is this necessary, since they are, of course, due to symmetry, structured in the same way, and the identifiers for each state within the sub-levels can be easily inferred.

Rolling flex map of secondary sub-level S

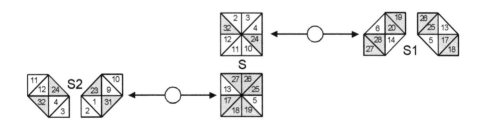

David Mitchell / Silverflexagons and the Flexatube

Map of secondary sub-level connections to home states 1A and 2A

Note that states S and T appear twice on this map. This map could equally well have been drawn with home state 2A in the centre.

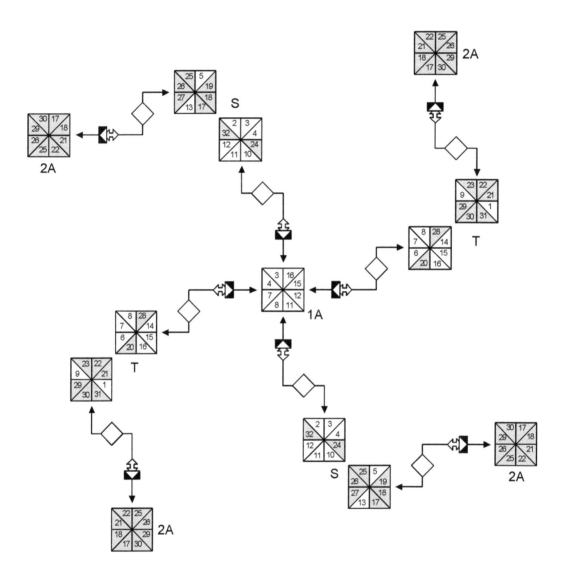

Map of secondary sub-level connections to home states 1B and 2B

Note that states U and V appear twice on this map. This map could equally well have been drawn with home state 2B in the centre.

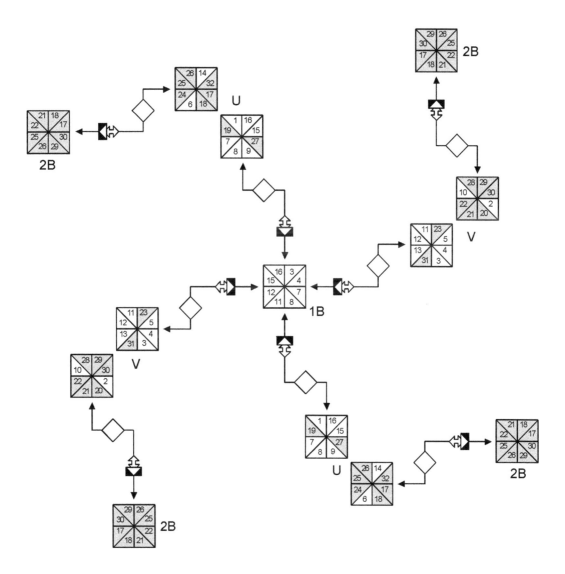

David Mitchell / Silverflexagons and the Flexatube

Accessing the tertiary sub-levels

There are eight tertiary sub-levels. The square states belonging to each of these form intermediate points on the routes between the main levels of the flexagon. The tertiary sub-levels can be accessed by a complex single tuck and inside swivel flex in the way shown here.

46.
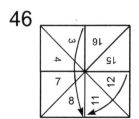

46. Begin with state 1A. Fold 12 onto 11 and 4/3 onto 7/8.

47.
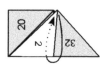

47. Fold flap 2/32 downwards in front then swivel it up inside the remaining layers.

48.
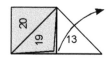

48. Fold 13 upwards to the right, allowing 19/20 to follow across.

49.
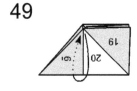

49. Repeat step 47 on flap 6/20. 19 will pull down as you begin to make this fold.

50.

50. Fold 23 upwards to the left.

51.
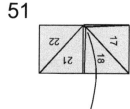

51. Separate the layers at the top and fold 18 downwards.

52.
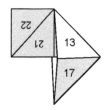

52. Turn over sideways.

53.
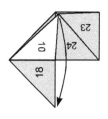

53. Separate the layers at the top and fold 10/24 downwards.

54.
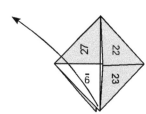

54. Separate the layers and fold 9/27 upwards to the left.

55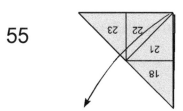

56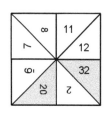

55. Fold the top layers diagonally downwards to the left.

56. This is state CC.

To return to the main levels simply follow these instructions in reverse.

Map of tertiary sub-level connections to home states 1A and 2A

This map could equally well have been drawn with home state 2A in the centre.

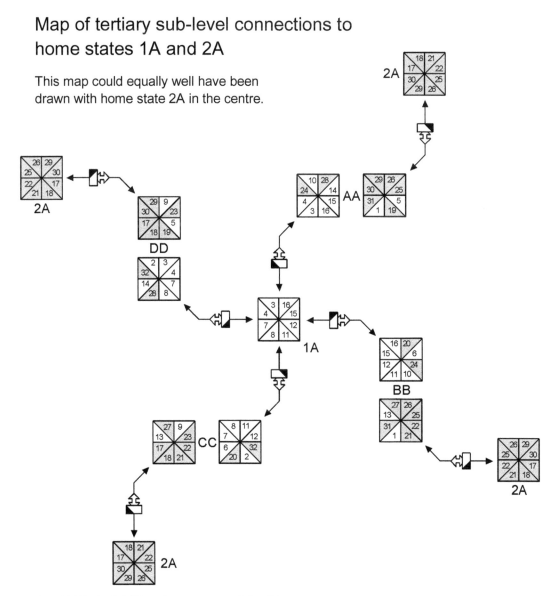

David Mitchell / Silverflexagons and the Flexatube

Map of tertiary sub-level connections to home states 1B and 2B

This map could equally well have been drawn with home state 2B in the centre.

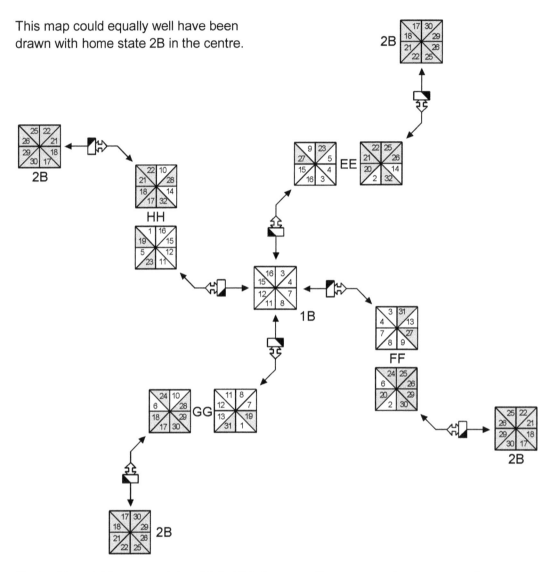

Each of the eight square states AA to HH shown on these maps gives access, via a single swivel flex, to a sub-level consisting of two quadrilateral states. Only one map of one of these eight sub-levels has been drawn.

Taken altogether the tertiary sub-levels of the Labyrinth Silverflexagon contain 8 square states and 16 quadrilateral ones.

Single swivel flex map of tertiary sub-level AA

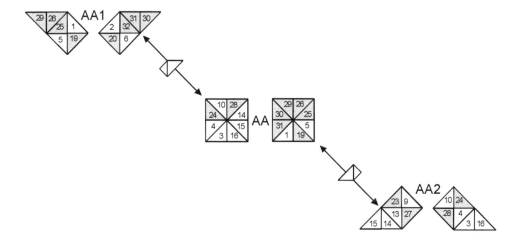

And finally

We are now in a position to enumerate the states of the Labyrinth Silverflexagon.

Including intermediate states (but not those belonging to sub-levels) main levels 1 and 2 consist of 4 square states, 16 hexagonal states, 48 pentagonal states and 4 oblong states.

The primary sub-levels consist of 16 square states, 32 hexagonal states and 64 pentagonal ones.

The secondary sub-levels consist of 4 square states and 8 hexagonal states.

The tertiary sub-levels consist of 8 square states and 16 quadrilateral states.

Adding all these together gives totals of 30 square, 56 hexagonal, 122 pentagonal, 4 oblong and 16 quadrilateral states, giving a grand total of states for the Labyrinth Silverflexagon of 228 states of all shapes, unless that is we have somehow missed finding a few along the way.

Given the complexity of the Labyrinth Silverflexagon that would not, perhaps, be at all unlikely.

The Flexatube

The Flexatube

The Flexatube is undoubtedly the greatest paperfolding puzzle ever invented. According to Martin Gardner it was first discovered in 1939 by a young mathematician called Arthur Stone while he was experimenting with flexagons. The Flexatube is not itself a flexagon, though it shares many similar properties.

The Flexatube is an open ended square section tube with square faces, each of which is divided into four segments by diagonal creases. These diagonal creases, and the upright ones where the faces meet, act as hinges between the segments so that the tube can be folded into various different shapes, and ultimately turned completely inside out. The challenge is to try to find out how this can be done using just the creases that are present when the folding starts.

Diagrams for seven and a half separate solutions are included in this book (on pages 96 to 113), or perhaps for six and a half solutions if you consider the Easy Street and Concealed Entrance solutions to be the same. I leave this judgement up to you.

The solutions fall into three categories. In some of the solutions, Central Line, Easy Street, Concealed Entrance and Deviation, the transformation can be achieved without bending or distorting the planes of the individual segments of the paper between the creases. In others, Around the Houses and Twin Peaks, only minimal distortion takes place. Finally, there are those other solutions, Threading the Needle, Reversing the Thread and the Gordian Knot, in which the paper is happily and completely distorted as the folds are made.

As you work your way through the solutions you will find that one particular state of the puzzle, which I call the Oxford Circus position, occurs in many of the solutions. It is therefore possible to multiply the possible number of solutions even further by combining the many different ways Oxford Circus can be reached and departed from.

A template for the Flexatube puzzle can be found on page 132.

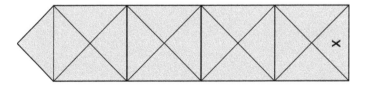

1. Cut the template out and crease carefully along all the boundaries between the segments. Fold all the creases backwards as well as forwards so that the segments move freely in both directions. Turn over sideways.

2

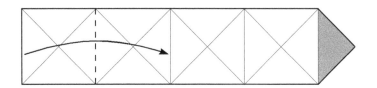

2. Fold the left edge inwards as shown. Cover the triangular tab at the right (shown shaded) with glue.

3

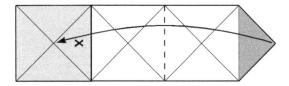

3. Fold the right end of the template inwards as shown and glue the tab on top of segment x. The result will be a squashed square section tube. Allow to dry then open the tube up to look like picture 4 below.

4

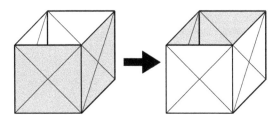

4. The challenge is to turn this tube completely inside out just by folding along the existing diagonal and upright creases and without making any new creases in the process.

Compound Flexatubes

Flexatube Stack puzzles

Flexatube Stack puzzles are created by stacking Flexatubes on top of each other. I have not been able to succeed in turning either uncut stacks, or stacks where a slit is cut to separate just one of the four edges between the Flexatubes, inside out. However it is possible to succeed in this where the slit separates two adjacent edges of the Flexatubes.

Templates and solutions are provided for double and triple stacks. You may like to experiment with larger stacks for yourself.

The Double Flexatube Stack puzzle

A template for this puzzle can be found on page 134. The solution is given on pages 116 and 117.

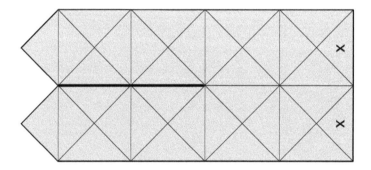

1. Cut the template out, then cut along the slit marked with a thick black line to partly separate the two Flexatubes. Construct the puzzle by following the instructions in steps 1, 2 and 3 of the Flexatube puzzle on pages 86 and 87.

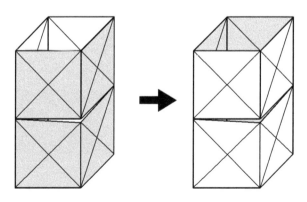

2. The result should look like this. The objective is the same as for the basic Flexatube puzzle i.e. to turn both tubes completely inside out just by folding along the existing diagonal and upright creases and without making any new creases as you do so.

The Triple Flexatube Stack puzzle

A template for this puzzle can be found on page 135. The solution is given on pages 118 to 120.

1

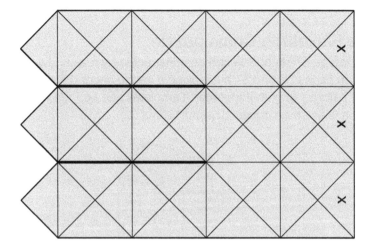

1. Cut the template out then cut along the slits marked with thick black lines to partly separate the three Flexatubes. Construct the puzzle by following the instructions in steps 1, 2 and 3 of the Flexatube puzzle on pages 86 and 87.

2

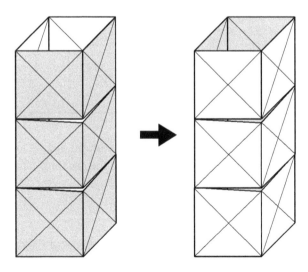

2. The result should look like this. The objective is the same as for the basic Flexatube puzzle i.e. to turn all three tubes completely inside out just by folding along the existing diagonal and upright creases and without making any new creases as you do so. The crux of the problem is, of course, to find a way to turn the middle Flexatube inside out. Once you have done that the rest is easy.

Flexatube Chain puzzles

Flexatube Chain puzzles are created by linking Flexatubes into chains but without otherwise glueing them together.

The Double Flexatube Chain puzzle

The Double Flexatube Chain puzzle can be constructed by making two Flexatubes in the way shown on pages 86 and 87 but making sure the second Flexatube is linked through the first before you glue it together. Templates can be found on pages 132 and 133. The solution is given on pages 121 and 122.

1

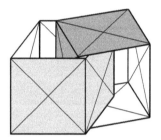

1. The resulting puzzle should look like this.

2

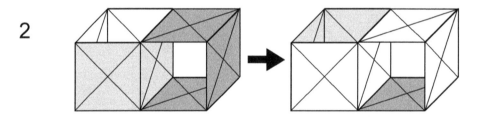

2. The objective is the same as for the basic Flexatube puzzle i.e. to turn both tubes completely inside out just by folding along the existing diagonal and upright creases and without making any new creases as you do so.

The Triple Flexatube Chain puzzle

The Triple Flexatube Chain puzzle is made by adding an extra Flexatube to the chain like this. The solution is given on pages 122 and 123.

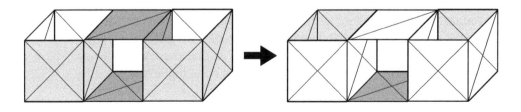

The crux of the problem is, of course, to find a way to turn the middle Flexatube inside out.

Once you have worked out how to solve the Triple Flexatube Chain puzzle you will also be able to solve Flexatube Chain puzzles of any length.

The Flexamat puzzle

The Flexamat puzzle is made by linking four Flexatubes together into a mat. This makes the puzzle appear, at first sight, almost impossible to solve. The solution is given on pages 124 and 125.

1

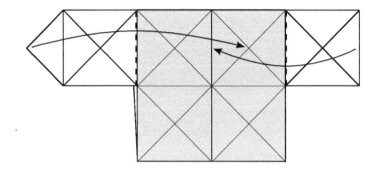

1. Begin by placing two complete flattened Flexatubes side by side and threading a third Flexatube through them both like this. Glue the third Flexatube together.

2

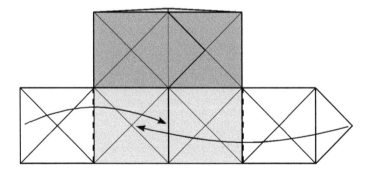

2. Thread a fourth Flexatube through the first two Flexatubes and below the third like this. Glue the fourth Flexatube together.

3

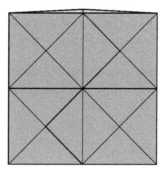

32. The result should look like this. The objective is the same as for all the other Flexatube puzzles i.e. to be able to turn all the Flexatubes completely inside out just by folding along the existing diagonal and upright creases and without making any new creases as you do so. Since the puzzle is symmetrical, being successful with one Flexatube will show you how to succeed with the others as well.

Solutions to the Flexatube

Solutions to the Flexatube

Central Line Solution

1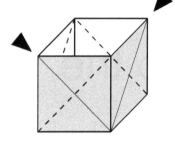

1. Collapse flat.

2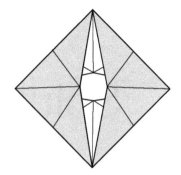

2. This is the collapse in progress.

3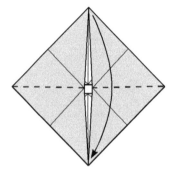

3. Fold in half downwards.

4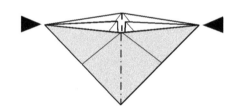

4. Squash sideways.

5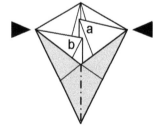

5. Continue squashing. Make sure that flap a goes backwards and flap b comes forwards as you do this.

6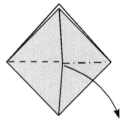

6. Once the squash is complete, rotate the front of the puzzle sideways by 90 degrees so that it looks like this then open out the front flap as shown.

7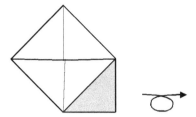

8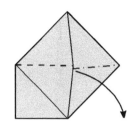

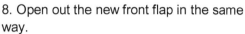

7. Turn over sideways.

8. Open out the new front flap in the same way.

9

10

9. Squash flat sideways.

10. Arrange to looks like this, then open out the front flap downwards.

11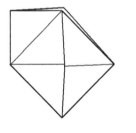

12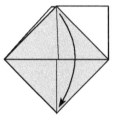

11. Turn over sideways.

12. Open out the new front flap in the same way.

13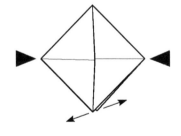

14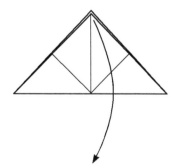

13. Separate the layers at the bottom and squash flat sideways.

14. Open out the front layers downwards.

David Mitchell / Silverflexagons and the Flexatube

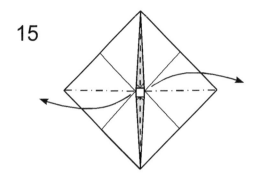

15. Lift the front layers upwards and outwards to reform the Flexatube.

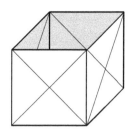

16. The Central Line solution has been achieved.

Easy Street Solution

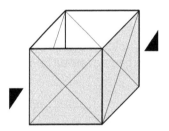

1. Squash flat.

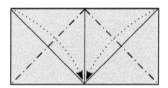

2. Fold both top corners backwards.

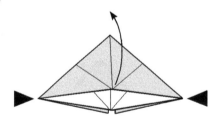

3. Squash slightly to open the centre.

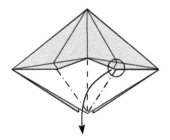

4. Reach inside the open mouth of the puzzle, hook the tip of your finger around behind the lower flap at the point circled. The flap will pull towards you and fold downwards.

5 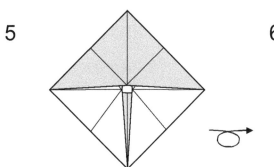 6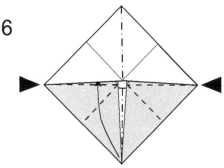

5. This is the Oxford Circus position. Turn over sideways.

6. Squash slightly to open up the centre and fold the bottom flap up inside the front layer.

7 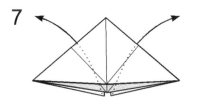 8 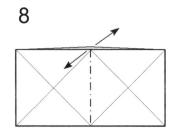 9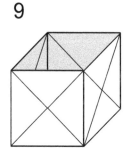

7. Bring the flaps at the back into sight.

8. Pull the layers apart.

9. The Easy Street solution has been achieved.

Concealed Entrance Solution

1 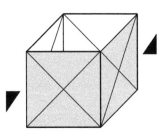 2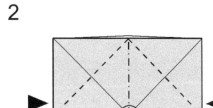

1. Squash the tube flat.

2. Apply gentle pressure to bring the two bottom corners together, making sure that the point marked with a circle rises up towards you. The corresponding point at the back should move away from you at the same time.

3

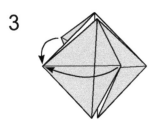

3. Fold the flaps at the front and the back to the left hand side.

4

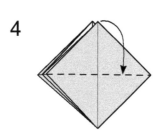

4. Fold the front flap down at 90 degrees.

5

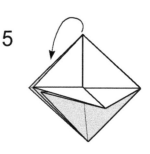

5. Fold the rear flap down at 90 degrees as well.

6

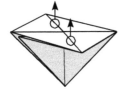

6. Insert your fingers gently underneath the flaps marked with circles and lift upwards.

7

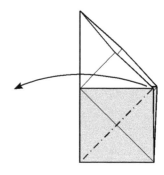

7. Fold the front layers to the left.

8

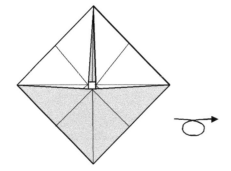

8. This is the Oxford Circus position. Turn over sideways.

9

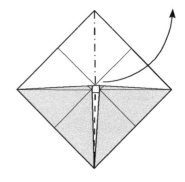

9. Lift up the centre of the front layer and squash the puzzle flat sideways.

100   David Mitchell / Silverflexagons and the Flexatube

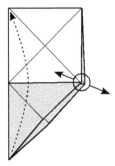

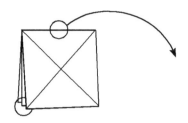

10. Separate the layers at the point marked with a circle and fold the bottom point upwards in between the other layers.

11. Take hold of the middle bottom left layer and the top edge at the points marked with circles and pull the top edge to the right.

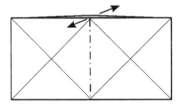

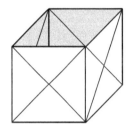

12. Pull the layers apart.

13. The Concealed Entrance solution has been achieved.

Deviation Solution

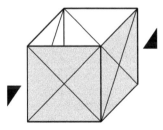
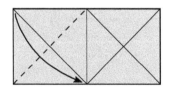

1. Squash the tube flat.

2. Fold the top left corner downwards in front.

3

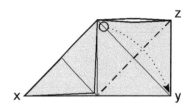

3. Fold the flap marked with a circle down inside the model. The puzzle becomes three-dimensional as you do this.

4

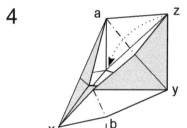

4. Fold point z behind then gently collapse the puzzle flat sideways while allowing point b to move towards you.

5

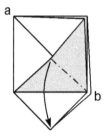

5. Pull the centre of the front layer downwards.

6

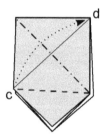

6. Fold the flap marked with a circle up inside the model so that c ends up at d.

7

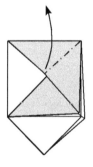

7. Pull the centre of the front layer upwards.

8

8. Turn over forwards.

9

9. Fold the front layer in half diagonally to the left.

10

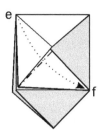

10. Fold the tip of the central layers at point e down inside the puzzle so that it ends up at f.

11

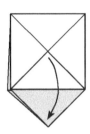

11. Fold the front flap downwards like this.

12

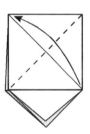

12. Fold the bottom right corner of the front layers diagonally upwards to the left.

13

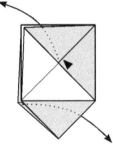

13. Apply pressure to the internal edge so that it pushes out upwards to the left. At the same time open the back bottom left corner out to the right. The puzzle will become three-dimensional as you do this and should look like picture 14.

14

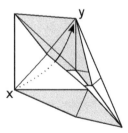

14. Fold the internal flap at x to y and flatten the puzzle.

15.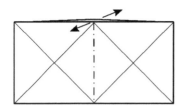

15. Open up the layers.

16.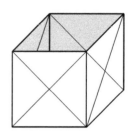

16. The Deviation solution has been achieved.

Around the Houses Solution

1.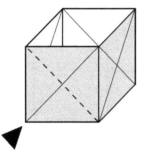

1. Push in one corner and flatten.

2.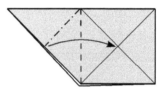

2. Bring the left hand flap forwards and squash symmetrically.

3.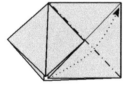

3. Fold the flap marked with a circle up inside the layers.

4.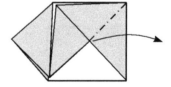

4. Pull the centre of the front layer to the right.

5.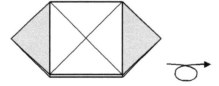

5. Turn over sideways

6.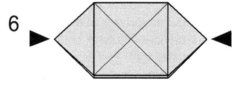

6. Squash to make three-dimensional.

7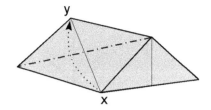

7. Take corner x inside the model to end up at y.

8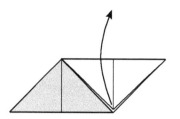

8. Open up the front flap.

9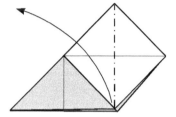

9. Fold the front layers upwards to the left.

10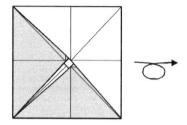

10. This is the Oxford Circus position. Turn over forwards.

11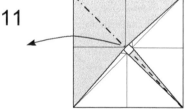

11. Pull the centre of the front layers to the left.

12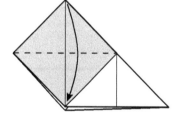

12. Fold the top flap downwards.

13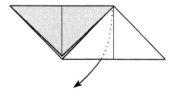

13. Reach inside the layers and pull out the internal flap.

14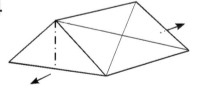

14. This is what the result should look like. Pull out both ends to collapse flat.

15. Turn over sideways

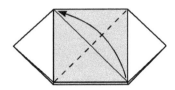

16. Fold the front layers in half diagonally upwards to the left.

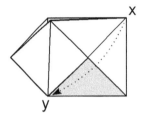

17. Reach inside the layers and fold the internal flap at x to y.

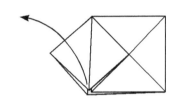

18. Pull the bottom point of the front layer to the left.

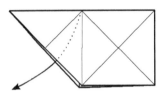

19. Open up the layers and pull out the internal flap.

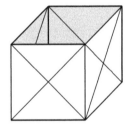

20. The Around the Houses solution has been achieved.

Twin Peaks Solution

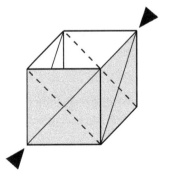

1. Push in two opposite corners simultaneously. As you do this the two original outer edges of the tube will meet each other inside. Push each slightly to one side so that they slide past each other until the puzzle collapses flat.

2

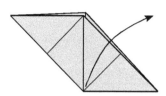

2. Open up the front layers so that the result looks like picture 3.

3

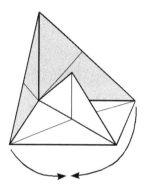

3. Pull both the outside corners downwards and inwards until they meet.

4

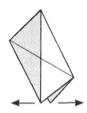

4. Pull the centre apart and flatten so that the result looks like picture 5.

5

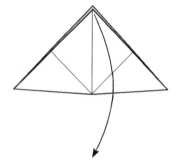

5. Fold the front layers downwards.

6

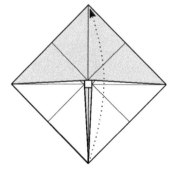

6. This is the Oxford Circus position. Fold the bottom point upwards behind.

7

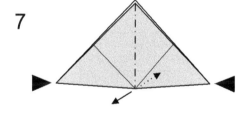

7. Open the centre and squash flat sideways.

David Mitchell / Silverflexagons and the Flexatube

8.

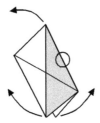

8. Hold the puzzle at the point marked with a circle then separate the bottom points and swing them apart and upwards so that the result looks like picture 9.

9.

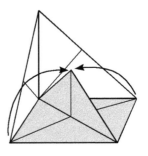

9. Bring the outside points together in front.

10.

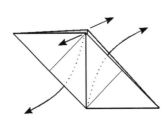

10. Open out the layers and bring out the internal flaps.

11.

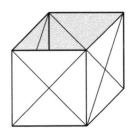

11. The Twin Peaks solution has been achieved.

Threading the Needle Solution

1.

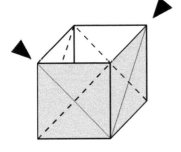

1. Collapse flat like this.

2.

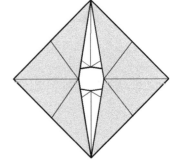

2. Collapse in progress.

3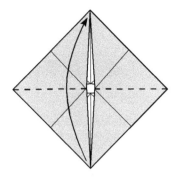

3. Fold in half upwards.

4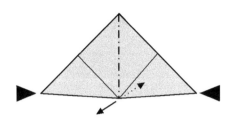

4. Squash sideways making sure both internal flaps move forwards as you do so.

5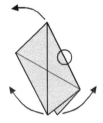

5. Take hold of the thinner side of the puzzle with your right hand at the point marked with a circle and use your left hand to pull the thicker layers upwards and to the left, separating the layers at the centre of the bottom of the puzzle as you do so.

6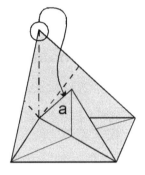

6. There is a slit in the back of flap A. Curve the circled point downwards and insert it inside this slit as far as it will go.

7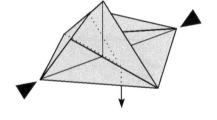

7. The dotted line indicates the position of this point inside flap A. Squeeze the two sides of the model gently as shown, then grasp the tip of the point from below and pull the point downwards.

8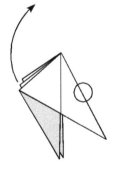

8. Hold the puzzle with your left hand at the point marked with a circle and use your right hand to pull the middle layers at the left upwards.

David Mitchell / Silverflexagons and the Flexatube

9

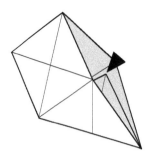

9. Push down on the middle flap to turn it inside out and complete the puzzle.

10

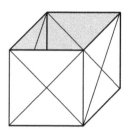

10. You have succeeded in Threading the Needle.

Reversing the Thread

1

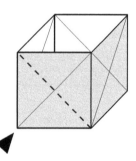

1. Push one corner in and flatten.

2

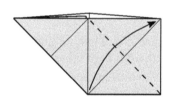

2. Fold the bottom left corner diagonally upwards as shown.

3

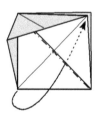

3. The puzzle will have become three-dimensional. Fold the new bottom left corner diagonally upwards behind the other layers to flatten the puzzle again.

4

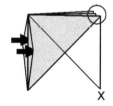

4. There are two pockets in the left hand edge of the puzzle. Insert the tip of one of your thumbs in each of these pockets and hold all the layers at the top right together with your second fingers. As you gently separate your thumbs point x will move upwards.

5

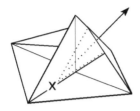

5. This is the result. Extract point x from the pocket.

6
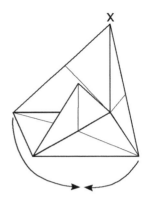

6. Bring the two outside corners together underneath to flatten the puzzle.

7

7. Pull the bottom layers apart and flatten sideways.

8
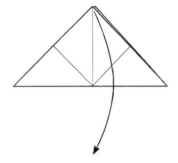

8. Fold the top corner of the front layers downwards.

9
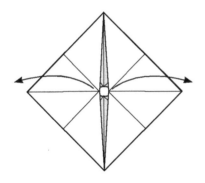

9. Open out the centre to reform the Flexatube.

10
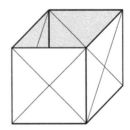

10. You have succeeded in Reversing the Thread.

David Mitchell / Silverflexagons and the Flexatube

The Gordian Knot Solution

1

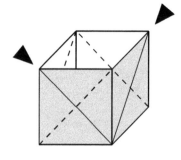

1. Collapse flat like this.

2

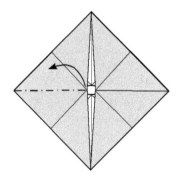

2. Lift the left front flap upwards slightly.

3

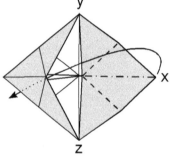

3. Lift up point x and fold it over and down through the central slit so that the tip just protrudes on the other side. As you do this points y and z will come together.

4

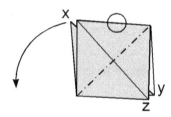

4. Hold the top edge at the point marked with a circle and pull point x firmly to the left.

5

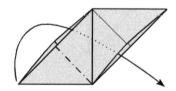

5. This is the Gordian Knot.

6

6. Fold the left hand point back through the slit in between the layers so that the result looks like picture 7.

7

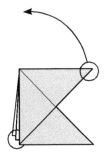

8

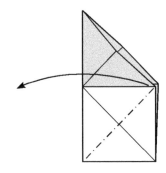

7. Take hold of the middle of the three flaps at the bottom left and pull the top right corner upwards.

8. Open the front layers out to the left.

9

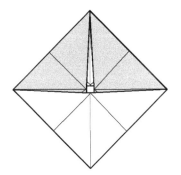

9. We have reached the Oxford Circus position yet again. I believe that completing this solution by turning the puzzle over and reversing the Gordian Knot route without damaging the paper is virtually impossible, but you are welcome to prove me wrong.

Solutions to the Compound Flexatubes

Solution to the Double Flexatube Stack puzzle

1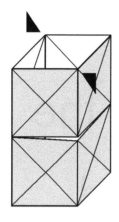

1. Flatten the puzzle so that the slit is at the front.

2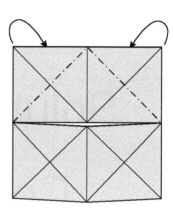

2. Fold both top corners backwards.

3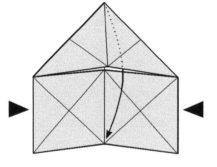

3. Squash slightly to open the pocket then reach inside to pull the internal layers out downwards.

4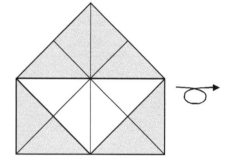

4. Turn over sideways.

5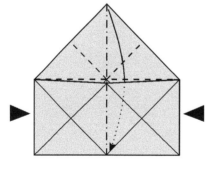

5. Squash slightly to open the pocket then fold the top point down inside.

6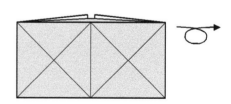

6. Turn over sideways.

David Mitchell / Silverflexagons and the Flexatube

7

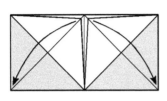

7. Fold the front flaps diagonally outwards.

8

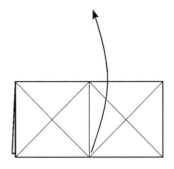

8. Fold the front layers upwards.

9

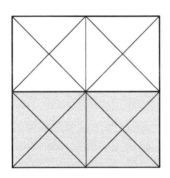

9. The first Flexatube has been turned inside out.

10

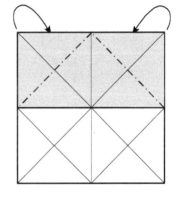

10. Rotate through 90 degrees and repeat steps 2 through 8 to turn the second Flexatube inside out as well.

11

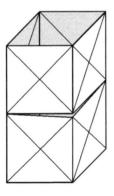

11. The Double Flexatube Stack puzzle has been solved.

Solution to the Triple Flexatube Stack puzzle

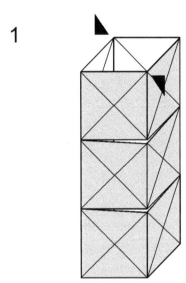

1. Flatten the puzzle so that the slits are at the front.

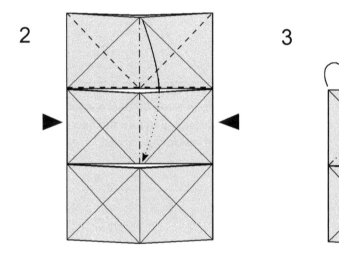

2. Squash slightly so that the front layer of the middle Flexatube moves forward then fold all the layers of the top Flexatube down inside the middle Flexatube as shown.

3. Fold both top corners backwards.

4

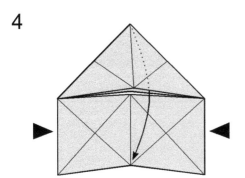

4. Squash slightly to open the pocket then reach inside to pull the internal layers out downwards.

5

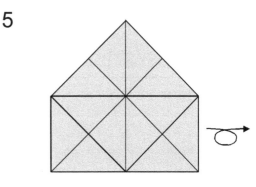

5. Turn over sideways.

6

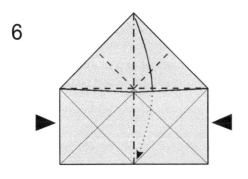

6. Squash slightly to open the pocket then fold the top point down inside.

7

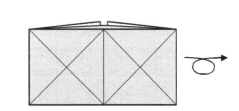

7. Turn over sideways.

8

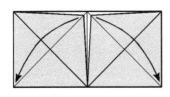

8. Open out the front flaps diagonally downwards.

9

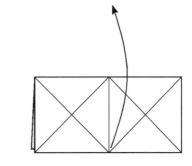

9. Fold the front layers upwards.

10

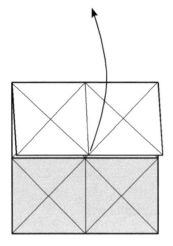

10. Fold the front layers upwards again.

11

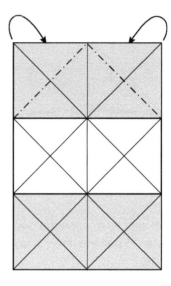

11. The middle Flexatube has been turned inside out. You can turn the top and bottom Flexatubes inside out as well by following steps 2 to 8 of the solution to the Double Flexatube Stack puzzle.

12

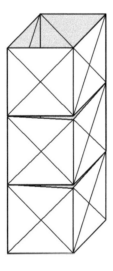

12. The Triple Stack Flexatube puzzle has been solved.

Solution to the Double Flexatube Chain puzzle

1

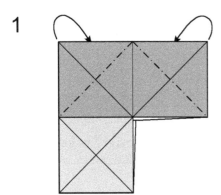

1. Flatten both Flexatubes and arrange like this. Fold both top corners backwards as shown.

2

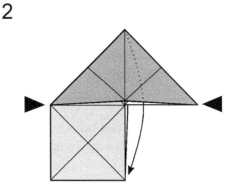

2. Squash slightly to open the pocket then reach inside and pull the internal layers out downwards.

3

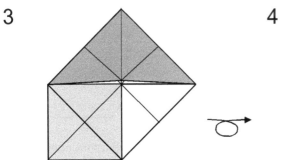

3. Turn over sideways.

4

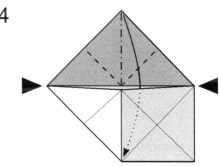

4. Squash slightly to open the pocket then fold the top point down inside.

5

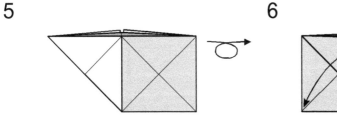

5. Turn over sideways.

6

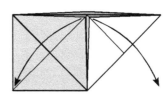

6. Open out both front flaps.

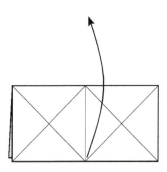

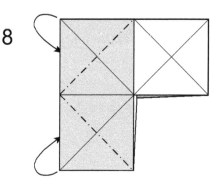

7. The first Flexatube has been turned inside out. Fold the front layers upwards.

8. Repeat steps 1 through 7 to turn the second Flexatube inside out as well.

Solution to the Triple Flexatube Chain puzzle

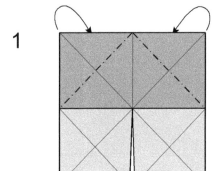

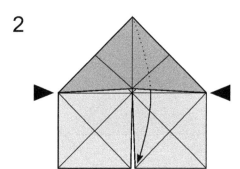

1. Flatten all three Flexatubes and arrange like this. Fold both top corners backwards as shown.

2. Squash slightly to open the pocket then reach inside and pull the internal layers out downwards.

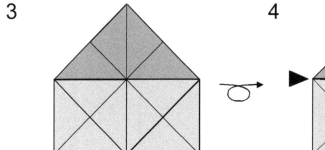

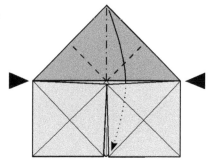

3. Turn over sideways.

4. Squash slightly to open the pocket then fold the top point down inside.

David Mitchell / Silverflexagons and the Flexatube

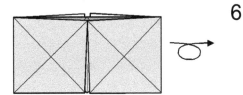

5. Turn over sideways.

6. Open out both front flaps.

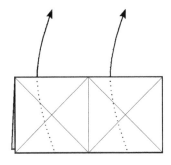

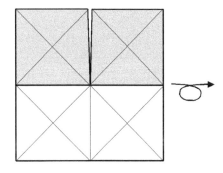

7. Fold the back layers upwards.

8. The middle Flexatube has been turned inside out. Turn over sideways.

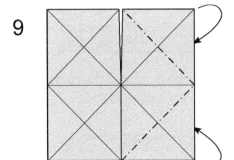

9. Repeat steps 1 through 7 to turn both the other Flexatubes inside out in turn.

Solution to the Fleximat puzzle

1

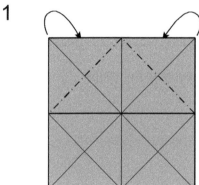

1. Fold both top corners backwards as shown.

2

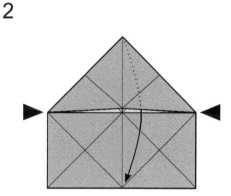

2. Squash slightly to open the pocket then reach inside and pull the internal layers out downwards.

3

3. Turn over sideways.

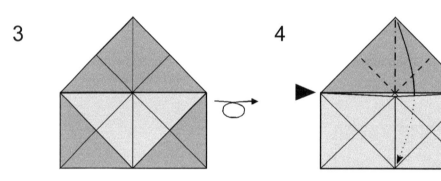

4

4. Squash slightly to open the pocket then fold the top point down inside.

5

5. Turn over sideways.

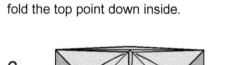

6

6. Open out both front flaps.

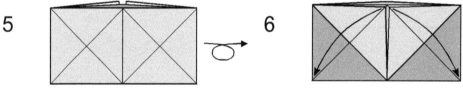

7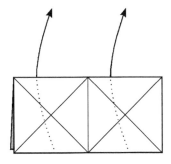

7. Fold the back layers upwards.

8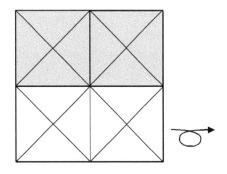

8. The first Flexatube has been turned inside out. Turn over sideways.

9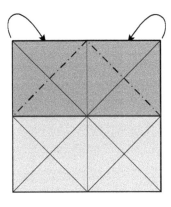

9. You can repeat steps 1 through 7 to turn each of the other three Flexatubes inside out in turn.

David Mitchell / Silverflexagons and the Flexatube

The Templates

The Zigzag Silverflexagon

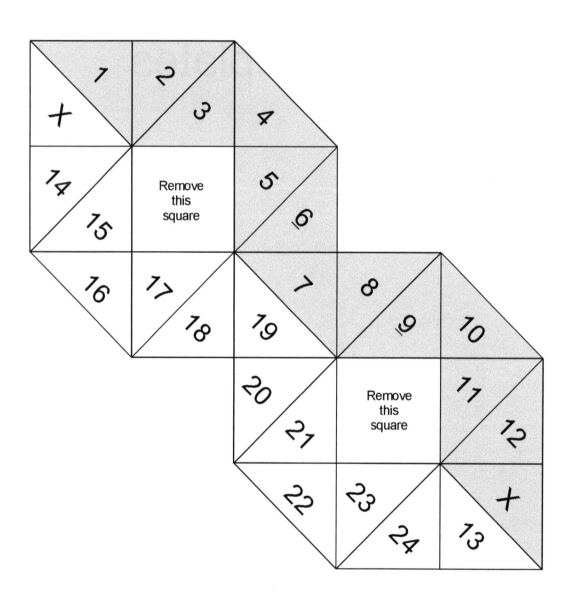

The Woven Flexatube

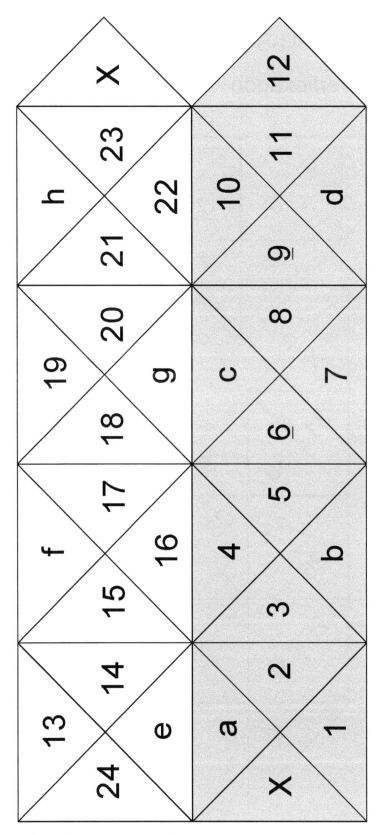

David Mitchell / Silverflexagons and the Flexatube

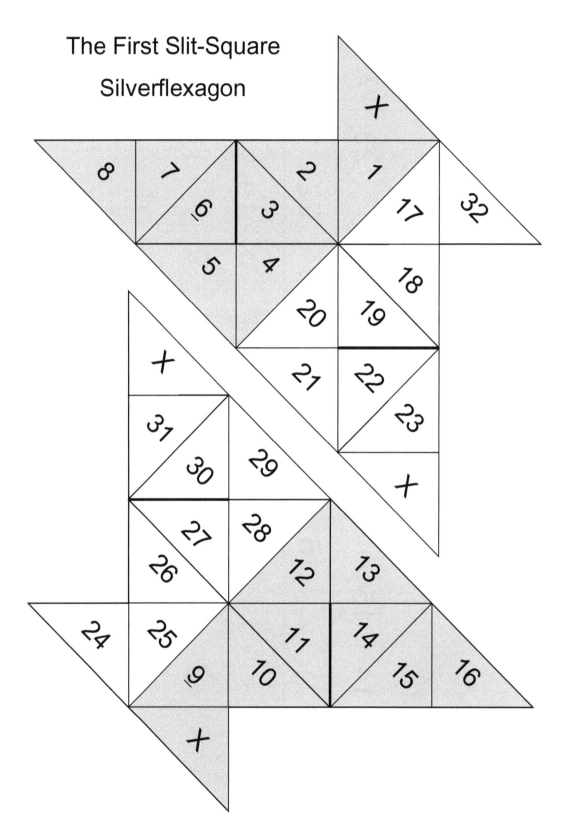

The Labyrinth Silverflexagon

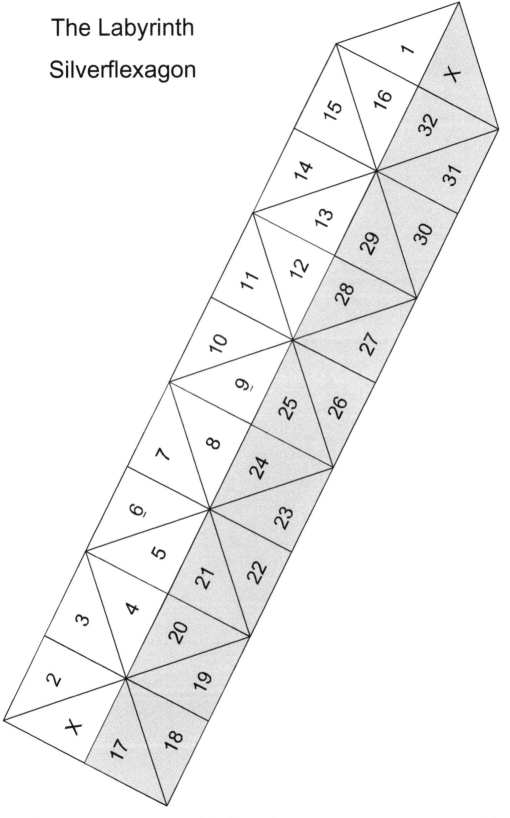

David Mitchell / Silverflexagons and the Flexatube

The Flexatube

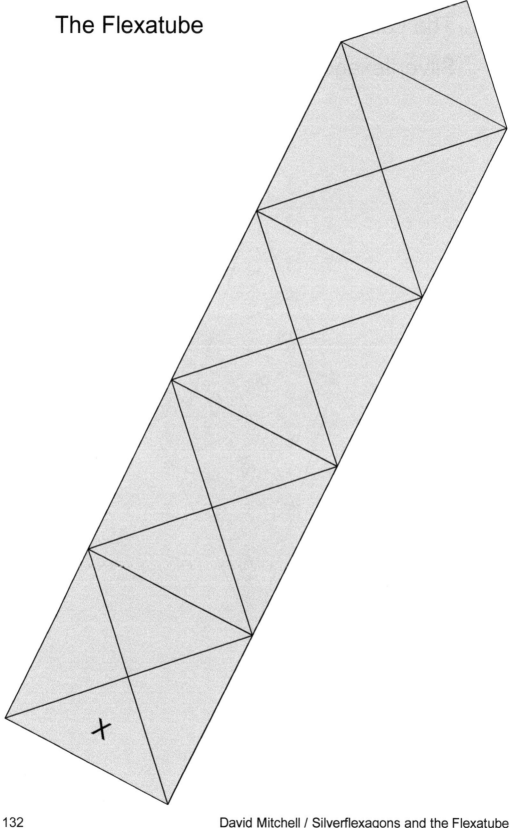

The Flexatube

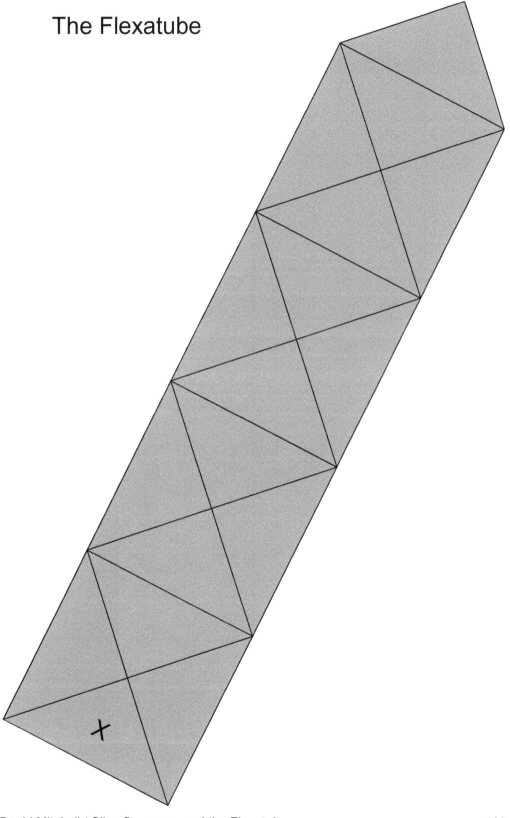

Double and Triple Flexatube Stacks

134   David Mitchell / Silverflexagons and the Flexatube

Lightning Source UK Ltd.
Milton Keynes UK
UKHW031842070221
378365UK00005B/113